The Holism-Reductionism Debate

The
Holism-Reductionism
Debate

In Physics, Genetics, Biology, Neuroscience, Ecology, and Sociology

By

Gerard M. Verschuuren

Table of Contents

Preface

This book is an introduction into the reductionism-holism debate, for aspiring as well as accomplished scientists. It is intended for those working in, or preparing for, research in any scientific field—ranging from the physical sciences to the life sciences to the behavioral sciences and the social sciences. It is certainly not meant for people specialized in areas dealing with the specific issue of reductionism in a strict philosophical sense; they won't learn much new from this book. In other words, this is not a monograph with specialized, original research, but rather an initiation into the debate—more like a textbook, if you will.

I use the term reduction in a rather neutral way—as a technique of doing research rather than an all-pervading worldview claiming that reduction is the only legitimate technique for doing research. I will also line up a reductionist approach with what is sometimes called a holistic approach. Very often holism is a term used interchangeably with, or at least related to, organicism, emergentism, interactionism, and systems theory. What these latter terms have in common is that they represent an approach that is supposed to be complementary to a reductionist approach, rather than an alternative.

In this book, I try to keep an impartial position regarding the discussion about reduction, reductionism, and holism by showing the pros and cons of positions others have taken. I will just examine various viewpoints in the discussion, and then let the reader decide what they are worth. The reader may sense sometimes where I stand myself, but I hope to remain unbiased and stay clear of any one-sided viewpoints as much as humanly possible. But because the reductionist approach usually does not need extensive backing, it may look like I spend more time in this book on the holistic side of the discussion, but that is only because the latter approach is not as prevalent in scientific circles as the former one and may therefore need more explanation.

Philosophers have the task of questioning and analyzing what most other people, including scientists, usually take for granted. For that reason, this book is basically a plea against any form of dogmatism in science, in favor of a more open-minded approach. Dogmas do not belong in science, although they do occur in the scientific community.

I decided not to add notes with references to my sources and citations, for the simple reason that I want this to be an introductory book for beginning as well as accomplished scientists. At the end of each chapter, though, I do mention a few references providing some background information for further personal study. The selection is limited and inevitably subjective.

A special thank-you I would like to extend to those who expressed certain insights better than I could ever on my own. In particular, I want to mention Stephen M. Barr, John Dupré, John Haldane, Leon R. Kass, and Michael A. Simon—to name just a few. They may even be responsible for certain wordings in this book, but obviously, they are not responsible for the final outcome; if I erred, it is entirely my doing. They and many others make me realize that originality only consists in the capacity of forgetting where your information came from.

I.   Reduction and Reductionism

The idea of reduction—not to be confused with the chemical term reduction, which stands for the gain of electrons by an atom or molecule—was first introduced by René Descartes in Part V of his Discourses of 1637, where he argued that the world is like a machine, its pieces like clockwork mechanisms, and that the machine could be understood by taking its pieces apart, studying them, and then putting them back together to see the larger picture again. Descartes' work was expanded by Newton (1643–1727) and ultimately culminated in the *Principia Mathematica* in 1687—one of the most influential science books ever written—in which Newton further advanced the idea of a "clockwork universe."

Soon it became common practice to apply this technique to any kind of scientific study—not only in the physical sciences, but also in the life sciences, and even in the social sciences. The technique became so popular and so successful that more and more scientists began to believe that reduction is the only way of understanding their object of study. This is when the discussion shifted from reduction to reductionism. Whereas reduction is merely a method or technique, reductionism is more of a creed or doctrine. If there are other tools than reduction, then the exclusive claim of reductionism cannot be true.

To someone such as Richard Dawkins, who says "Reductionism is one of those things, like sin, that is only mentioned by people who are against it"—one could reply that this may be true for reductionism, but the technique of reduction in and of itself deserves a much more serious scrutiny, instead of declaring it either a "virtue" or a "sin."

1. Simplified Models

The Nobel Laureate and physiologist Peter Medawar used to describe scientific research as "the art of the soluble." It is the art of devising hypotheses that can then be tested by practicable experiments. This is typically done by "reducing" the complexity and multiplicity of events to a manageable *model* related to an analyzable problem. This idea has inspired some among us to claim that science has in essence a reductionist approach. Whether that is really the case remains to be seen, and also depends to a large extent on how we define reduction and reductionism.

Let us find out first why science does work with models, and actually cannot be without them. Obviously, all events and phenomena that science deals with are usually too complex to be studied in their entirety. Inquiry into *all* of the possible causes of a phenomenon—say, a certain disease—would be a virtually unending process. Any possible variable would have to be taken into consideration: the temperature, the air pressure, the position of the stars, and an infinite number of other factors. Every variable, even the most unlikely ones, might eventually turn out to be relevant. Nevertheless, we are prone to designate some factors as irrelevant. The reason for this is that within the setting of our "research program," we expect certain factors to have no effect; or sometimes we are just not interested in their effects.

In a basic version, research may start out with a simple model of, say, two variables. The independent variable is the one freely chosen by the scientist in order to trace its influence on the other, so-called dependent, variable. With the help of these terms, we are able to define a scientific research "problem" in a more specific way as follows: What is the influence of the independent variable on the dependent variable? However, apart from the two variables studied in an experiment, there is an unlimited group of so-called non-experimental variables which may potentially interfere with the relationship between the dependent and the independent variable. The nightmare of all scientists is that their research is going to be blown by some non-experimental variable. Hence, it is of great importance to eliminate the influence of these non-experimental variables in advance and to the greatest degree possible. This does not hold for the irrelevant variables, as they were proven by preceding experiments to have no influence at all—provided these experiments were correctly done. In principle, however, any other variable may be a potential source of interference. Some are known to interfere—the relevant variables—others we do not yet know of—the potential variables.

The consequence of this fact is that other (possibly) interfering variables have to be kept under control or are assumed to be constant, which leads to the so-called "ceteris paribus" clause. Control can be exercised in various ways. One way is to keep the interfering variable at a steady state (temperature, etc.), or to eliminate it if possible (light, gravity, etc.). In a new and unknown area of research, however, we often encounter a special problem, because we do not know ahead of time where interfering variables are to be located. In such cases, searching may become like "fishing in troubled waters." A solution to this problem can be found in using large samples, allowing the interfering variable(s) to exhibit a random distribution. Randomization is a common technique for ensuring a valid comparison of the groups involved.

Usually scientific research is a mine-field of "hidden" variables. In principle, research is aimed at relevant variables—framed in a question like "What is, as a consequence of this test, the change in color, temperature, direction, or movement?" Scientists tend to focus their attention on what is important for their central issue. So an essential part of a research program is limitation of the area to be investigated. Those who are considered good scientists can be identified by their capacity to reduce their many question marks to a manageable problem. Much of the genius of research workers lies in their capability of selecting what is worth observing.

The microbiologist and pathologist W. I. B. Beveridge once likened scientific research to "warfare against the unknown." He considered this a useful analogy because it suggests an important tactic. The procedure most likely to lead to an advance is that of concentrating one's forces on a very restricted sector chosen because the "enemy" is believed to be weakest there. Weak spots in the defense may be found by preliminary scouting or by tentative attacks. When stiff resistance is encountered, it is usually better to seek a way around it.

Apparently, "good" scientists are considered experts in defining the formulation of their problem. But there is more to it. It is also their job to remove the process or object under investigation from its natural context. Expressed technically: During the experiment, interfering variables have to be ruled out or controlled. Therefore, scientists limit themselves to a simple setting, often done in a laboratory. To put this in an image, an experiment needs to take place in the "test-tube-like shelter" of a laboratory, removed from the complexity of nature. In a test tube (*in vitro*), for instance, a biological experiment is easier to keep under control than inside the body (*in vivo*).

A visit to an immunological lab, for example, shows us immune cells in test tubes attacking tumor cells or producing antibodies all by themselves. This may create the impression that the immunity system is an isolated self-regulating system. It is only after broadening their scope that immunologists discovered that the immune system interacts to a considerable extent with other bodily systems. Extrapolation from a simple model to a complex situation is at best debatable and at worst misleading.

In short, a problem needs not only to be demarcated, but also to be manageable as well. Every research object has to be simplified in the form of a manageable model that can be both defined and controlled. Because the study of complex systems calls for simplifying models, these models have a limited scope and will only hold when the assumed boundary conditions are satisfied. As Monica Anderson points out, "When studying a predominately Reductionist discipline, such as physics, you often see textbook phrases like 'All else being constant...' or 'In a closed system...' which indicate that some degree of Reduction has already been made, so that what follows in the problem statement is all that matters." This indicates that some degree of model simplification has already been made based on some kind of "ceteris paribus" clause.

This way, the model helps us study the behavior of any system under ideal or abstract conditions. This requires approximation or idealization. To understand the motion of a billiard ball, for instance, it makes sense to treat the ball as if it were a perfect sphere of infinite rigidity, without worrying about its chemical composition, scratches on its surface, and all sorts of other details. This kind of idealization does not alter the fact that ideal gases, perfect conductors, prototype mammals, Hardy-Weinberg populations, or utopian societies do not really exist—they only do in models. This makes Fulvio Mazzocchi conclude that "the strategy of putting the object of study in an artificial, controlled and simplified experimental situation, even if it is done for heuristic purposes, is not neutral and has its price—it severs the essential link between the object and the natural context to which it belongs."

A simple example of the use of models is a common household tool that displays the current air pressure in the atmosphere—or in a fancier version, a history of a sequence of recent air pressures. In either case, the tool uses a simple model to predict whether we are in for sun, clouds, or rain. It is obvious that such a model is quite limited. But even if many more variables or factors are added to the model, the predictability of the weather leaves much to be desired. There is always a "ceteris paribus" clause involved to the effect of "everything else being constant." A model used in weather forecasting is never a complete replica of the weather system, and is therefore limited in its prediction capabilities. Some say that even the movements of tiny butterfly wings may have a ripple effect on the weather.

What is closely related to this discussion on models is the causality issue—the question of what the "cause" is that makes something happen. Some consider the real cause of something as a *sufficient* condition, in whose presence the effect is bound to occur. For instance, scientists who want to make certain plants develop in a certain way look for a sufficient condition in whose presence that development does occur. In doing so, they may hit upon a certain fertilizer, a certain degree of humidity, and a certain insect transporting pollens; and by combining these conditions, they make sure that the plants in the field will develop as desired.

Usually, however, scientists slim the "cause" down to one of the antecedent conditions—a *necessary* condition—to the exclusion of other conditions. First of all, this is a practical procedure. If we want to eliminate the cause of a plant disease, we look for a necessary condition in whose absence the disease cannot occur. And thus we may find a certain mold that causes the disease; next the mold can be killed so that the disease will disappear. But also under less practical circumstances, scientists tend to select one, or at most a few, conditions as "the" cause. In this view, causes are "abnormal" conditions; they are not complete sets of causal factors, but factors that "make the difference." All the other conditions are then called "background conditions," which are not considered the focus of investigation. Apparently, the model we are using determines what the background conditions are.

Take the causes of embryonic development. Usually, geneticists take genetic factors as the causes of development, for it is these factors that create specific differences between organisms. This is not to deny that there are also non-genetic factors involved, such as food or oxygen, which are also necessary conditions in development. However, in most cases these latter factors do not make a specific difference between organisms; hence, geneticists consider them merely as background conditions, without denying their causal impact. As a matter of fact, every explanation assumes there are certain background conditions which can be left unspecified; they are certainly necessary conditions but can be left out of the equation for the sake of simplicity.

No matter how we look at it, we end up with simplifying models which have a limited scope and hold only when the assumed boundary conditions are satisfied. All scientists use models, whether they realize it or not; they can even use different models to describe the same underlying reality. As Sean Carroll puts it, "Sometimes you might want to talk about a box of gas as a fluid with pressure and velocity, other times you might want to talk about it in terms of atoms and molecules." Models focus on what is relevant by omitting what is considered irrelevant. Obviously, models of this kind, like model airplanes, are simplifications of reality. They do not replace or replicate reality. It is a truism that a "super-model" honoring all the details of our complex world would be as complex as the world itself.

Because of this, models provide only representations of *aspects* of reality—perhaps more or less relevant, but always partial. This way, they provide partial insight into the complexities of the real world. But we should always be aware that looking at an organism, for instance, as if it were a machine does not make it a machine. There is a constant danger of being seduced by our models, of being so enchanted by the fact that they fit the data that we ignore the fact that countless other models might equally well fit our observations. Models never fully replicate what they represent.

This conclusion may also have some consequences for the way we look at the so-called fundamental "laws of nature." These laws are understood to be universal in scope, meaning that they apply to everything there is in the universe. But then the question arises: What role do laws of nature play in science if it is models that represent what is happening in the world? One possible response is to argue that laws of nature govern entities and processes in a model rather than in the "real" world. Fundamental laws, on this approach, do not state facts about the world but hold true of entities and processes in the model. Yet, there is a natural tendency among scientists, when they have a successful scientific model, to attempt to apply it to as many problems as possible. Jacques Monod said somewhere in the fifties that what holds for *Escherichia coli* holds for an elephant as well. But it is also in the nature of models that these extended applications are potentially dangerous. As the British philosopher of science John Dupré puts it, "The abstractions that work well in one context may eliminate what is essential in another."

The core message here is that models never fully replicate the original of which they are a "replica." To show that a model is never a complete replica of its original, the example of aggression may be helpful. Aggressive behavior in human societies is a phenomenon that has been "modeled" or "mapped" in many different ways, given the fact that all science is abstraction. A current *biological* version has it that aggression is programmed in the genes and is part of human nature. Such is the way our brains supposedly function, for this behavior has been influenced by our genes, and in turn these genes are the outcome of the evolutionary process our ancestors have gone through. On the other hand, from a *sociological* viewpoint, aggressive and violent human behavior can be understood as a consequence of environmental factors in society. Aggressive behavior is supposed to have been acquired in an aggressive environment, perhaps by mimicking aggressive examples from the immediate vicinity or from the TV-screen. Different again is the *psychological* viewpoint which makes psychologists rather search for personal issues such as frustration as a source of aggression. Anyone who has been frustrated in satisfying their own needs or in achieving a set goal is assumed to react in an aggressive way.

The outcome is that we have at least three different models "mapping" the same phenomenon, aggression, in three different ways. Put differently, each model depends on "how you look at it," for all science is abstraction. Hence the distinction between the original and its model is important; once we mix them up, models may be mistaken for the original. Looking at an organism as if it were a machine does not make it a machine. It is completely acceptable to state, for instance, that we can make a DNA-map of a human being; but it is quite something else—and one might even say something illegitimate—to declare that human beings are identical with their DNA-maps.

What is more, even a collection of numerous models together would never exhaust the original itself. As each model has a limited scope, we may wish to integrate many of them in order to insure better coverage of the original. However, this would leave us with a staggering number of possibilities for making combined models. Wim Van Der Steen has argued that simple arithmetic suffices to show that integration of models easily becomes a self-defeating aim. Suppose we start out with ten models and are content with pairwise combinations. We would end up with forty-five models over and above the original ones. Surely the need for integration would be enhanced, not diminished by this. Plus we would be left with the problem of how to connect them. Besides, we may seriously question whether this type of procedure would deepen our understanding. Yet, science keeps expanding by offering us more and more diverse models. It is probably still true that 90% of all the scientists who have ever lived are alive today. And they are not idle! So they make models multiply at a high speed.

The hallmark of models—demarcation and simplification— relates not only to the details of research, but also to its general scope. Put in very general terms, psychologists watch different things than biologists, and biologists look for other issues than physicists—just to name a few types of scientists. Scientists are like poachers who use the "conceptual jacklights" of their models to catch "hares"; obviously, this would make other "animals" elude notice. As a consequence, each scientific field or discipline creates its own "world," its own phenomena, its own facts, thus abstracting from other viewpoints.

Seen this way, a psychologist has an "eye" for psychological phenomena, whereas a biologist perceives only biological facts. The very same event can be looked at from different "angles" or "perspectives," with different "glasses," within different frames of reference—and therefore, with different models. Different perspectives can, in fact, complement each other and allow scientists to acquire a more complete knowledge of a certain phenomenon. This caused someone like the late neurologist J. H. van den Berg to say that the designation H_2O betrays the manifold varieties of the element water. Drink H_2O, if you can. Swim in H_2O, if you want. Is water not H_2O then? Of course it is, but water is not just H_2O—it is *also* H_2O. As he puts it, "H_2O is only one aspect of water-ness." All science is per definition partial and fragmentary

In a nutshell, scientific modelling is a scientific activity, the aim of which is to make a particular part or feature of the world easier to understand, define, quantify, visualize, or simulate. It requires us to select and identify relevant aspects of a situation in the real world so we can then use different types of models for capturing those different aspects. Hence, every model has its limitations. A case in point is Newtonian physics, which is highly useful except for the very small, the very fast, and the very massive phenomena of the universe. Every model also has its own "point of view." Sometimes you might want to talk about a box of gas as a fluid with pressure and velocity, other times you might want to talk about it in terms of atoms and molecules. Again, every model leaves out certain aspects and variables. Astronomers can ignore quantum fuzziness when calculating the motions of planets and stars. Conversely, quantum chemists can safely ignore gravitational attraction between individual atoms because this force is about 40 powers of ten feebler than the electrical forces between them.

There is much more to models, but that is beyond the scope of this book. Let us conclude this discussion as follows. Although models in science are very helpful, and actually even inevitable, Caltech physicist and engineer Carver Mead, a guiding force in three generations of Silicon Valley technology, was right when he paradoxically said: "The simplest model of the galaxy is the galaxy."

For further reading:

Bunge, Mario. *Method, Model and Matter*. Dordrecht: Reidel, 1973.

Frigg, Roman, and Stephan Hartmann. "Models in Science," *The Stanford Encyclopedia of Philosophy*, Edward N. Zalta (ed.),   2012 http://plato.stanford.edu/archives/fall2012/entries/models-science/.

Hesse, Mary. *Models and Analogies in Science*. Notre Dame, IN: University of Notre Dame Press, 1966.

Keller, E. F. Making Sense of Life: Explaining Biological Development with Models, Metaphors, and Machines. Cambridge, MA: Harvard University Press, 2002.

2. Levels of Organization

The various sciences are sometimes likened to different levels of a tall building—particle physics on the ground floor, then the rest of physics, then chemistry, and so forth, all the way up to psychology and the social sciences on the top floor. There is a corresponding hierarchy of complexity—atoms, molecules, cells, organisms, and so forth. At first sight, the analogy of a building sounds neat, but it has serious drawbacks, as we will see. It may be an intuitive comparison, but it is also dangerous and may lead to serious debates.

Nonetheless, it is very common among scientists—especially in the life sciences, the behavioral sciences, and the social sciences—to distinguish various levels of organization in a so-called hierarchy of life, also called an *organizational hierarchy*. Usually this hierarchy is seen as based on reality, not just on a mere arbitrary scheme of classification. Each level in the hierarchy is seen as composed of building blocks from "lower" levels, but in itself is also "building material" for "higher" levels. The relation between these levels is what they call a "part-whole" relationship. Again, it sounds very intuitive, but these terms may be very ambiguous. Take the following example. Cells are made of molecules, but they are not made of genes. So unlike genes, molecules are the building blocks of cells. Genes, on the other hand, are parts or components of cells, as cells do contain genes but are not made of genes—they are made of molecules. Does it really matter? Sometimes it does.

In addition, what counts as a part, and which parts are basic, are matters best settled in a particular context of enquiry. Atoms would count as basic parts of hydrogen if it is oxidized to form water, but not if it is converted into helium by a thermonuclear reaction. And then there is another caveat. A brick, for instance, has certain mass, which is a property of the whole brick, whereas none of its parts has that mass. However, aggregates of this sort do not provide interesting cases of a whole-part relationship; even a dedicated reductionist can comfortably accommodate them, since the brick's mass results simply from summing the mass of each part. So the whole-part concept usually needs further exploration before we can speak of a genuine part-whole relationship.

This may sound like hairsplitting, but there are some important implications for when we want to distinguish levels in the hierarchy. A rather detailed and extensive division from "low" to "high" could be as follows: atoms | molecules | organelles | cells | tissues | organs | organisms | populations. This list can be read this way: atoms are parts of molecules, molecules are parts of organelles, organelles are parts of cells, cells are parts of tissues, tissues are parts of organs, organs are parts of organisms, and organisms are parts of populations. This construction is called a "hierarchy"—not so much because the science of a "higher" level is really "on top of" the previous one, but because entities at each level can be understood in either direction: in terms of their components (i.e. at a "lower" level) and in terms of what they themselves compose (at a "higher" level). So atoms are parts of molecules, but also the other way around, molecules are composed of atoms. In addition, other types of hierarchy have been suggested: subatomic particles form atoms, atoms form molecules, molecules form monomers, monomers form polymers, polymers form cells, cells form organs, organs form organisms, organisms form societies, societies form economies, etc.

Thinking this way may indeed be a very neat way of conceiving of the world and structuring scientific research. But misconceptions might easily slip in. Sometimes, for instance, life scientists talk about the hierarchy of life without making explicit that they have introduced a new kind of level that, on further analysis, does not belong in there. As mentioned earlier, some like to refer to the level of genes, but that is a misleading way of talking, for genes may be parts of cells, but cells are not composed of genes. Since cells are just not made of genes, genes do not make for their own level in the hierarchy, given a certain definition of a part-whole relationship. It might be more appropriate to place genes on the level of molecules.

When getting to the highest levels, even more confusion may set in. Is the species level in biology, for instance, a legitimate level in a hierarchy? It depends. If it is the top level above populations, it could be appropriate, for a species does consist of populations, and populations are the parts of a species. But it would not make for a legitimate level below any other higher level. Ecosystems, for instance, do not consist of species but of populations. On the other hand, something that could be positioned below ecosystems would a biocoenosis, which is a community of interacting populations from various species (see chapter 13).

What this discussion shows us is that organizational hierarchies can be rather diverse. Besides, it can be very limited; some areas of research, for instance,—such as paleontology and systematics—have even been ignored in this discussion about biological hierarchies. It is not immediately clear where they would belong in an organizational hierarchy. No wonder, the discussion about organizational levels can at times become very murky. Therefore, more analysis would be necessary when it comes to organizational levels. But let's leave that work up to others.

Notwithstanding all these ambiguities, nowadays scientists consider themselves specialized not so much in the *object* of their research as in the *level* of their research. By this we mean that they prefer to work at a certain level inside a much larger hierarchy. They focus on what is specific for a certain level and thus abstract from other levels. Each level spans a specific range of phenomena. Sociology is less general in scope than biology; biology is less general than chemistry; and chemistry is less general than physics—whatever that actually means. Physical and chemical properties are present in anything there is, but biological properties are said to be present only in systems specifically recognized as biological objects, and sociological properties only in systems identified as sociological objects—therefore, applying them to anything else would be considered inappropriate. It is the possession of biological properties that makes something a "biological object," and something similar holds for a "sociological object."

It is safe to say that science has become increasingly specialized and compartmentalized. One could, for example, categorize the various sciences in terms of organizational levels, although it remains hard to do so consistently, as we found out:

- Communities: sociology, synecology.

- Species: evolutionary biology, systematics, taxonomy.

- Populations: ecology,   population   genetics, anthropology.

- Organisms: physiology, anatomy, psychology.

- Cells: cell biology, microbiology, neuroscience.

- Molecules: chemistry, biochemistry, molecular biology.

- Atoms: atomic   physics,   nuclear   physics,   plasma physics.

- Particles: particle physics, quantum mechanics.

As a consequence of such a categorization, there are also fields that try to combine and integrate various levels. These are called interdisciplinary fields, which partially have become sciences of their own: astrophysics, biophysics, chemical physics, econo-physics, geophysics, medical physics, physical chemistry, and so on.

What is noteworthy about hierarchical levels, though, is that each level in the "hierarchy" has its own characteristics. Each level may have properties not shown at lower levels. This phenomenon is called *emergence*. The properties of water molecules, for instance, do not show up at the level of their components, the hydrogen and oxygen atoms. Something similar can be said about biology. Sleeping is a characteristic of organisms, not of cells. An organism may also undergo an aging process, which in turn is impossible for a population; what a population does undergo is an evolutionary process, whereas an organism does not. All levels—and thus their corresponding scientific fields—are characterized by their own specific properties, and thus by their own distinct concepts and explanations.

Here is another example from biology to show what the existence of levels in a hierarchy entails. Ventilation, respiration, oxygenation, and oxidation are very closely related processes but they take place at different levels: Ventilation occurs at the level of organisms, oxygenation at the level of tissues, cellular respiration at the level of cells, and oxidation at the level of molecules. Here are the differences:

- Ventilation is the movement of air into and out of the lungs whereas respiration is the exchange of oxygen and carbon dioxide. Ventilation facilitates the process of respiration; without ventilation, respiration cannot occur. The ventilation mainly involves the lungs while the respiration mainly involves the respiratory surfaces, including alveoli and blood capillary walls.

- Respiration, in physiology, is defined as the transport of oxygen from the outside air to the cells within tissues, and the transport of carbon dioxide in the opposite direction. The physiological definition of respiration should not be confused with the biochemical definition of respiration, which refers to cellular respiration.

- Oxygenation occurs when oxygen molecules enter the tissues of the body. For example, blood is oxygenated in the lungs, where oxygen molecules travel from the air and into the blood. At low partial pressures of oxygen, most hemoglobin is deoxygenated. At around 90% (the value varies according to the clinical context) oxygen saturation increases according to an oxygen-hemoglobin dissociation curve and approaches 100% at partial oxygen pressures of >10 kPa.

- Cellular respiration: the metabolic process by which an organism obtains energy by reacting oxygen with glucose to give water, carbon dioxide, and ATP (energy). Although physiologic respiration is necessary to sustain cellular respiration and thus life in animals, the processes are distinct: cellular respiration takes place in individual cells of the organism, while physiologic respiration concerns the bulk flow and transport of metabolites between the organism and the external environment.

- Oxidation takes place inside the cell, when hydrogen electrons are passed down a "respiratory chain" of electron-transport compounds, located in the cell's mitochondria, until they reach oxygen and form water.

Each one of the above concepts refers to its own level, forcing us to keep our terminology as precise as can be. This is even true of a concept as vague as "life." Whereas it is possible to say that a cell is alive, it is not really appropriate to say that a molecule is alive. Even a virus is really nothing but a package of molecules that cannot give evidence of life without the surroundings and the organization provided by a cell. The cell is considered to be the elementary unit of life. As a consequence, all living organisms are cellular, either unicellular or multicellular.

What is said here about biology in particular could also be applied to all the other sciences. One could make the case that all sciences deal with hunks of stuff, albeit of different sizes and different natures—stars, stones, cells, animals, human beings, molecules, atoms, particles, and so forth. All science is *abstraction*, and therefore partial and fragmentary. Physics and biology, for instance, each clearly operate on separate levels in the organizational hierarchy of life. The physical sciences operate at the level of atoms and sub-atomic particles, whereas the life sciences, the behavioral sciences, and the social sciences are usually focused on higher levels. Not only does this lead to differences in concepts and phenomena, but also to a more fundamental difference in approach. We could characterize these differences with the following core concepts and keywords: causality | functionality | emotionality | cognitivity. They make for different "commitments" on the part of scientists.

1. The search for causality is based on a commitment to searching for causes and their effects. It is a principle that guides all sciences and occurs at all levels of the hierarchy, including the level of the physical sciences. It can easily be formalized in simple statements with the following structure: *If X, then Y* (but there is much more to it, of course). In more general terms, the principle of causality states that like causes always have like effects. It works at any organizational level (although there is a vivid discussion going on in the field of quantum mechanics whether it also holds at the sub-atomic level; see chapter 7). Causality is the link, for instance, between heartbeats and blood circulation, for it is the heartbeats that keep the blood circulation going.

It is worth noting that the principle of causality is a basic and fundamental presupposition of doing research. It is the assumption that nothing happens without a cause—which is a presupposition that cannot be proved or disproved (see chapter 17). One could argue that it is logically impossible to prove that something has no cause at all, since searches like these never reveal the *absence* of their object. Causality can never be conclusively defeated by experiments since causality is the very foundation of experiments. Since science can never prove there is causality in this universe, it must *assume* there is. The tool of falsification, for instance, is actually based on this very assumption as well: The fact that scientific evidence can refute a scientific hypothesis is only possible if like causes do have like effects. Without this principle, there would not be any falsifying evidence. When we do find falsifying evidence, we do not take this as proof that like causes do *not* always have like effects, but as an indication that there is something wrong with the specific case of causality we had conjectured up in our minds. So the principle of causality cannot be refuted, not even by falsification.

2. The search for functionality is based on a commitment to searching for causes too, but specifically for those causes that have *successful* effects—which is typically the case in the life sciences. So the principle of functionality is not at odds with the principle of causality. Functional relationships are always causal, but causal ones are not always functional. The relationship between heart beats and heart sounds is certainly causal—beats produce sounds—but it is not a functional relationship—since the sounds do not benefit the survival of the organism. The relationship between heart beats and blood circulation, on the other hand, is not only causal—the heart beats make the blood circulate—but also functional, for the beats have a successful effect—the beats are there "in order to" have the blood circulate, which enhances the chances of survival and reproduction. Scientists in search of functionality are more interested in a mechanism's success than in its causes. It is a special feature of the life sciences that they assume the existence of functions—which are chains of cause and effect that are successful in enhancing survival and reproduction. The principle of functionality makes scientists search for successful effects, whatever their causes may be.

Whereas causality works at all organizational levels, functionality is limited to the levels where the life science, the behavioral sciences, and social sciences are operational. This leads us to the following question: What is the causality behind this functionality? How did functional features come to be? What is the causality, for instance, behind the functional green color of the caterpillars of the white cabbage butterfly? Since Charles Darwin's theory of natural selection, the answer would be that they are green, as a population, because of two different types of causes. On the one hand, they are green individually due to some particular genetic mechanism—which is sometimes called a "proximate" cause, usually based on mutations (see chapter 10). On the other hand, they are green collectively, as a population, because of the fact that their green ancestors successfully deceived potential predators in the past—which is often called an "ultimate" cause (see chapter 11). The green color is a cause that makes for a functional biological design and hence will be promoted by natural selection.

It is the latter type of causes that is important in the search for functionality. Selective reproduction is the causal explanation of how the green color of certain caterpillars could become so widespread in the entire population or species. The green color "made" it in evolution among these animals, because it is a successful and functional design in the environment they live in. Natural selection selects what is functional by favoring causes that have "successful" effects. So functionality is somehow causality tested in the furnace of natural selection, generation after generation. Functionality of causes that have successful effects is as basic to the life sciences as the principle of causality is to all the natural sciences.

This explains why causality does, but functionality does not, occur in the physical sciences, for the latter do not deal with selection based on survival and reproduction. When the temperature rises, gases do not expand "in order to" keep the pressure constant. There is no functionality here. But hearts do beat "in order to" keep the blood flow going. A beating heart is a good functional design, and natural selection favors this cause because it has a "successful" effect. So the causal relationship between heartbeats and blood circulation is functional, whereas the causal relationship between heart-beats and heart-sounds is not, since the latter connection does not contribute to the success of an organism in survival and reproduction.

3. The search for emotionality is based on a commitment to searching for a different kind of causes, namely those found in the "internal state" of animals and human beings. It is a principle that can operate at the level of the life sciences, the behavioral sciences, as well as the social sciences. It looks for motivations, incentives, drives, feelings, emotions, fears, passions, and mood changes—sometimes collectively addressed as instinct in animal behavior studies and ethology. Some have defined emotionality as the articulation of a natural response to a stimulus built on instinct. It is a measure of an organism's emotional reactivity to a stimulus.

The study of animal behavior, for instance, deals with the fact that organisms do not always react to a stimulus as a "reflex machine" would do based on a simple cause-and-effect chain. Certain kinds of stimulus-response behavior are dependent on motivation, which is a designation of the internal state of an animal. Of course, it is possible to study motivational systems such as sex drives and hunger drives in terms of their causal mechanisms, but motivation as such is a new principle in the field of animal behavior (ethology), helpful in explaining phenomena such as displacement activities and redirected activities, all of which reveal something about the performer's state of motivation. Motivations are a prelude to action. Feelings about food, companions, etc., can be measured by making animals work to achieve it. The harder they are willing to work, the more "it matters" to them, the more they are motivated and driven to act.

Not only ethologists, but also sociologists and psychologists may base their approach on the fundamental presupposition of emotionality. Sigmund Freud's libido, or sex drive, derives from it. Emotionality can sometimes overrule rationality. Emotionality is used here as a generic term for various forms of emotional energy. All of these can be analyzed in terms of causal or functional mechanisms, but they are not a new kind of cause in itself; they just have a different status. It is in this context that drives, feelings, motives, emotions, and temperaments have obtained their own status. They further specify the conditions under which a cause-and-effect chain comes "into effect."

4. The search for cognitivity is based on a commitment to searching for causes that are of an "intentional" nature. Some

call this intentionality, but the term intentionality may easily be misunderstood (see chapter 16). It is taken here in the sense the term was introduced by Jeremy Bentham in his doctrine of consciousness for the purpose of distinguishing acts that are intentional and acts that are not.

What makes cognitivity different from emotionality? The drives, impulses, instincts, or motives that we find in animals are usually directly or indirectly related to sex or food. (No wonder, behavioral experiments with animals usually enforce behavior with food rewards.) They are related to the principle of emotionality. But they are very different from believes, reasons, or intentions that can be found in human beings. In this view, beliefs and thoughts are not just another kind of cause, but are an entity on their own, connected with the principle of cognitivity in the mental realm (see chapters 14 and 16). "Thinking" and "willing" are considered mental activities belonging to a category which is distinct from the category of activities such as "eating" and "running." Bodily activities are causally and physically related—running makes you eat. Mental activities, however, are also causally, but not physically, related; thinking of "two to the power of two" makes you think of "four," provided you have had some training in logic and mathematics. There is causality involved, but it does not seem to be of a physical nature. The same could be said when beliefs and intentions make people act in a certain way.

Are "intentions" and "beliefs" acceptable entities in science, particularly in the behavioral sciences and the social sciences? There is quite some discussion as to whether intentions do exist, and if so, whether they also occur in the animal world, or only in the human world. Some argue that intentional idiom can, or even should, be replaced either by behavioristic language or by the language of neuroscience. But then we enter the debate on reductionism already, so let us keep that discussion for later. For now, let it suffice to state that behavioral and social scientists usually assume that intentions and beliefs do play a causal role in human behavior. One could even make the case that we can only make sense of causality (#1) by first drawing on the experience of ourselves as being our own causal agencies.

Here is an example to clarify and support this assumption. When some person nudges you under the table, you have reason to assume there is more to this action than neurophysiological causes and the like. Apart from causes, functions, and emotions, there is an *intention* behind it. An intention is a thought or belief about something; it can be a belief about what the world is like, a belief about one's own motives and goals, or even a belief about someone else's believes. Intentions are different from motives, as they are cognitive and/or conscious. Blinking an eye, for instance, can be seen as a process of chemical causes and biological functions, but winking at someone seems to be different from blinking an eye—it is an intentional act related to (cognitive) intentions and beliefs. That's where the principle of cognitivity separates from the principle of emotionality.

Can we prove or corroborate the principle of cognitivity? Again the answer is: No, we cannot. It is an assumption that can never be falsified; a search for cognitivity can never reveal the *absence* of its object. As a matter of fact, it reveals its very presence for the following reason. Any discussion about assumptions cannot possibly be done without invoking the principle of cognitivity, for assumptions are necessarily of a cognitive nature. Nonetheless, cognitivity has been a very controversial issue. We will get back to this later on (see chapter 13).

If we apply the four distinctions made above to a phenomenon such as aggression again, we could come up with the following. First, in terms of *causality*, aggression is caused by genes, hormones, and certain provocations. Second, in terms of *functionality*, it can be a functional response that enhances the chances of survival and reproduction. This is most obviously the case in terms of attacking a prey to obtain food, or in anti-predatory defense. It may also be the case in competition between members of the same species, if the average reward (e.g. status, access to resources, protection of self or kin) outweighs average costs (e.g. injury, exclusion from the group, death). Third, in terms of *emotionality*, aggression is highly correlated with certain emotional states. Stimulation of the amygdala, for instance, results in increased aggressive behavior in animals. Psychologists would say that so-called reactive aggression is usually highly emotional and can incite someone to a hate crime. Finally, in terms of *cognitivity*, aggression is under the influence of ideological beliefs. So-called proactive aggression such as racism or homicide is typically reasoned, and focused on carrying out some specific intention. In short, related to a phenomenon such as aggression, each one of these four approaches makes for its own distinct analysis.

To wrap up this chapter, we may conclude that the idea of organizational levels does carry some ambiguities and obscurities, yet it is a concept that most scientists have been using or are willing to use. It clarifies the way science is typically done as well as the way the various sciences distinguish themselves from each other. Even scientists who never use the world "level," often still think in terms of levels.

For further reading:

Capra, Fritjof. *The Web of Life: A New Scientific Understanding of Living Systems*. New York: Anchor Books, 1997.

Johnson, Steven Berlin. *Emergence: The Connected Lives of Ants, Brains, Cities, and Software*. New York: Scribner, 2001.

Pumain, Denise. *Hierarchy in Natural and Social Sciences*. New York: Springer-Verlag, 2006.

Strogatz, Steven H. *Sync: How Order Emerges From Chaos In the Universe, Nature, and Daily Life*. New York: Hyperion, 2004.

3. Interaction between Levels

Once we acknowledge some form of organizational hierarchy and the existence of properties that emerge at a higher level without existing at a lower level, the question arises as to *how* those properties emerge. It seems evident to most scientists that they do emerge at higher levels, but how?

Earlier we used the term *emergence* for the fact that each level has properties not shown at lower levels. A bee's nest, for instance, is the emergent result of the cooperative action of multitudes of individuals. But where do these new properties come from? Are they determined and/or caused by the level(s) below them and/or by the level(s) above them? In other words, what is it that causes emergence and what does emergence cause? How are we to account for the occurrence of properties on higher levels of organization when these properties are absent on a lower level?

According to reductionism, in its most common and general form, higher levels are fully determined by levels below them. This view hardly leaves any space for emergent properties—but that may not be quite true as we will see shortly. A contrast to general reductionism is the viewpoint of holism—sometimes also called emergentism, organicism, interactionism, or more recently, systems biology. Holism is the idea that entities at a higher level can have properties—so-called emergent properties—that are not explainable from the sum of their components at a lower level. The principle of holism was concisely summarized by Aristotle in his *Metaphysics* (1045a10): "The whole is more than the sum of its parts." Jan Christiaan Smuts later coined the term "holism" as "a tendency in nature to form wholes that are greater than the sum of the parts through creative evolution." We could formalize this idea as follows: $(1 + 1) > 2$. However, it should also be stated that, in the words of Fulvio Mazzocchi, "The whole is not only more than the sum of its parts, but also less than the sum of its parts because some properties of the parts can be inhibited by the organization of the whole."

Do we really need something like holism? There are some strong indications. Here are two simple cases where reductionism would seem insufficient. Consider the example of two resistors in a circuit. Whether they are connected in series or in parallel makes quite a difference for their total resistance. So the properties of this system as a whole are not given merely by a complete knowledge of the properties of the two individual parts. Additional information about the circuit diagram is needed. A similar statement could be made about the working of a protein. Information about the amino-acid composition of the protein is not enough to explain how the protein folds three-dimensionally before it can be operational. Again, additional information would be needed about the larger "system."

So the question remains: How could emergent properties ever arise from the properties of entities located at a lower level? There are various versions of emergence claims as distinguished by Franck Varenne. A *weak* and/or *nominal* version claims that wholes are dependent upon their lower-level constituents in the sense that the emergent properties do not apply to the underlying entities themselves. A *strong* version claims that emergent properties have power over the underlying entities—so they are not reducible to the properties of these entities. These macro-to-micro effects are termed downward causation.

Before we can analyze these claims, we need to delve into a rather philosophical, actually metaphysical, discussion: What are the basic ontological elements of our universe? The answer to this question could run along these lines: Our universe appears to be "stuffed" with things that are marked by properties, and that are mutually connected by relationships. In other words, there appear to be three different kinds of "entities" in this world. First, we have *things*, sometimes called substances, such as molecules, cells, organisms, species, communities, cultures, and the like. Next we have *properties*, making certain things inorganic and others organic, or certain creatures plant-like and others animal-like. Finally, we have *relationships* between things, such as the relationship bees have to their colonies, or football players to their team. When we say "Joe is tall and single," the word "single" does not designate a property like "tall"—instead it refers to a (missing) relationship.

These distinctions seem to be rather basic and yet vital in the way we tend to think and talk about this world; they go as far back as the philosopher Aristotle, and can still extensively be found in the philosophy of the late Harvard philosopher Alfred North Whitehead. They leave many questions open, though, and have been interpreted in very different ways. It is characteristic for philosophers to put question marks behind everything regular people consider standard. Some philosophers would claim, for instance, that substances do not exist but are merely bundles of properties; once we abstract from these properties, there is really nothing left, so they say— with Bertrand Russell being one of them. Some others think that relationships don't really exist—it's all a matter of substances and their properties—with Leibniz being one of them. Then there are some who do accept relationships; they may even treat substances as mere bundles of relationships, with properties being the mere outcome of these relationships —with Hegel being one of them. For those, though, who accept all three categories as real, in line with Aristotle—and probably with most of us—could come up with the following characterization of a whole-part relationship: The *properties* of a whole (a *substance*) will typically depend upon *relations* among its proper parts (*substances*) as well as on *properties* of the individual parts.

This discussion has important implications for science, and especially so for the sciences at higher organizational levels. Take, for instance, the problem as to how "lifeless" molecules could ever give rise to life. Many consider this an outdated debate that once raged between mechanists and vitalists. According to mechanists, the "whole" (e.g. a cell or an organism) is nothing more than the sum total of its parts; only the parts are "real," and the rest is fiction. But vitalists, for their part, would argue that a cell is certainly "more" than the collection of its molecules, or that an organism is "more" than the collection of its cells—or more in general, that the whole is "more" than the sum total of its parts—because of something, a substance or entity, that they called a "vital principle."

Another example of this debate in history is the antiquated theory of "life-force" in fermentation. In the early 1800s, vitalists argued that ferments—which we now refer to as enzymes—are linked to a living cell. Destroy a cell, they said, and ferments can no longer cause fermentation after the destruction of an immaterial "life-force." Then in 1896, the

German chemist Eduard Buchner mashed a group of cells with sand until they were totally destroyed. He then extracted the liquid that remained and added it to a sugar solution. He was amazed to discover that the cell-free liquid still caused fermentation, although it no longer carried the "life-force" deemed necessary to bring about fermentation. Apparently, the ferments themselves could cause fermentation, even though separate from any living organism. That sounded like a victory for mechanicists in their battle against vitalists, ending the claim that biology is materially discontinuous with the physico-chemical sciences. That discovery basically marked the end of vitalism.

However, one could argue that this controversy was spurious. Vitalists actually took the same philosophical stand as mechanicists by only focusing on substances and properties, without acknowledging the impact of relationships (the interactions) between all components. They both ignored that there could be "more" to life than substances and their properties, namely a "surplus" to be found in the relationships between the parts—that is, in the organization of the whole. In that sense, the whole is indeed composed of parts only, and yet it is "more" than the sum total of its parts. Nowadays, this view has become more and more popular under the heading "systems theory."

At the moment we take relationships into consideration as well, complexity still remains complex, but is not so "irreducible" anymore; complexity and emergence are just of a highly *interactive* nature. Interactionism takes relationships as very real entities. The atmosphere, for instance, is a veritable playground of emergence, which includes clouds, cold fronts, the Jetstream, hurricanes, tornadoes, and the list goes on and on—all of which are part of a complex network of relationships. To take this to an extreme, even the movements of tiny butterfly wings may have a ripple effect on the weather.

Most scientists nowadays would acknowledge the importance of relationships in the architecture of our universe. There are numerous examples to demonstrate this. The properties of oxygen (O_2) and hydrogen (H_2) do not account for the properties of water (H_2O); water owes its new properties to the relationships (or the interactions) between two atoms of hydrogen and one atom of oxygen in a water molecule. A similar explanation holds for the relationship between an enzyme and its cofactor; it is only when they are combined and can work together that they acquire a catalytic property. The enzyme alcohol dehydrogenase, for instance, requires the presence of zinc as a cofactor for it to work. Or take the chemical element carbon (C) with its rich potential of relationships that harbors an unusual polymer-forming ability (that is, within the temperature range commonly encountered on earth). This explains why all of organic chemistry is carbon-based, making carbon the "favorite" building block of the living world on planet Earth.

Most scientists would agree that it is hard to explain phenomena like these if we didn't acknowledge the impact of relationships in the universe—as well as in our philosophy. Take the parts of a table; they have the same weight, length, and surface volume before and after assembly, yet it is hard to deny that something new has come into existence after a table has been assembled. Or take this example from chemistry: When two or more (different) compounds share the same molecular formula but different structural formulas, they are called isomers. For instance, there are three different compounds with the molecular formula C_3H_8O: methoxyethane, propanol, and rubbing alcohol. These are different substances, with different chemical properties. Yet these differences are not based on differences in the composing atoms. It is rather the arrangement of those molecules that determines whether the substance will be methoxyethane, propanol, or rubbing alcohol.

Here is another case from biology: Cells are no longer seen as clumps of protoplasm. Instead they turned out to have a strong compartmentalization, which ensures that certain molecules such as enzymes are not only clustered in distinct cell compartments, but are also firmly confined to those compartments, thereby restricting reactions that might otherwise compete with one another. It is within the setting of larger compositions that components turn out to be capable of activities and phenomena which fail to show up in an isolated state. Understood this way, the whole is certainly "more" than the sum total of its parts, and yet there are no extra substances or entities involve, other than a complex underlying structure. Consequently, a fundamental tenet of *systems* biology is that cellular and organismal constituents are interconnected, so that their structure and dynamics must be examined in intact cells and organisms rather than as isolated parts.

In other words, we should not confuse holism, and its synonyms, with vitalism. Nowadays, vitalism is an outdated theory that once claimed there is some mysterious something about living things, exempt from physico-chemical laws. Vitalism is more like a perversely mystical "theory" of emergence. Instead, both reductionism and holism agree that living systems are made of the same kind of matter as non-living systems, and that they obey the same physical laws. They only differ as to the question if those laws state the sufficient as well as necessary conditions essential to the description and explanation of living systems.

If all of this is true, then whatever emerges during the process of evolution is not only the outcome of the substances involved and of their properties, but also of the new properties that come forth from the specific relationships between those components. Holism acknowledges the ability of complex systems to give rise to emergent novel properties that are not predictable from the examination of individual components. Emergent properties have to bring with them additional causal power, which cannot be extinguished by the causal power conferred by their physical bases—otherwise they would be merely epiphenomena.

A simple example is provided by the inability of detailed knowledge about the molecular structure of water to predict surface tension, which is a macroscopic phenomenon reflecting emergent behavior among water molecules. But the same holds for phenomena at higher levels. Gene networks and human social networks, for instance, may exhibit isomorphic patterns of self-organization due to their interactive structure, no matter whether genes or humans are the components. Thus, "self-organizing" phenomena occupy a special place in our discussion, because components are used to explain the behavior of the system, yet it is the nature of their interactions (not their specific characteristics) that generate patterns of new behavior, which are often referred to as emergent properties of the system.

Again, most emergent properties are effects at a higher organizational level which cannot be observed at a lower component level. The basic concepts of temperature and entropy, for instance, can be meaningfully applied only to systems of particles, but not to single particles. A single water molecule does not have a temperature since temperature is defined only for groups of molecules interacting with each other. And depending on the temperature, these water molecules will form water vapor, liquid water, or solid ice. These three states have very different properties. Could their different behaviors be predicted from the properties of individual molecules, such as van der Waals forces? It seems to be very unlikely, if not impossible. There are indeed cases, such as the kinetic theory of gases, where one can predict all the important properties of the system as a whole, given a good understanding of the components of the system. But in more complex systems, emergent properties of the system are said to be almost impossible to predict from knowledge based only on the parts of the system.

But holism goes even a step further. Not only does it acknowledge the emergence of novel properties in complex systems, it also "relaxes determinism in favour of recognizing unpredictability as intrinsic to complex systems," in the words of Fulvio Mazzocchi. This makes him suggest to replace "the idea of a deterministic (prescriptive) law by a non-prescriptive law that just provides a number of constraints. More than actually determining the development of organisms or other complex systems, such laws—if so intended—rather delimit the framework in which they occur. In other words, the 'law as a constraint' [...] defines a restricted field of possibilities, within which complex systems develop, but without imposing a unique way of undergoing this process."

For those who accept the existence of emergent properties—but again, not all scientists do—the question arises whether properties at a higher level can affect properties at a lower level in a way comparable to the way lower level properties have an effect on higher-level properties. Even those scientists who accept the idea of organizational levels—as most do—may not accept causation in either direction. Many of them tend to interpret organizational levels in a one-directional way of stepping "down" the hierarchical "ladder" as being the best, or even the only, way of doing science—no matter whether it is in the life sciences, the behavioral sciences, or in the social sciences. But some have reason to question this claim. Donald T. Campbell introduced the term "downward causation" for hierarchically organized systems. He defined it as follows: "[A]ll processes at the lower level of a hierarchy are restrained by and act in conformity to the laws of the higher level."

Whereas reductionism favors a *bottom-up* (or upward) causation, holism allows also for a *top-down* (or downward) causation—the whole (up at a higher level) determines to some degree the behavior of its parts (down at a lower level). To explain the idea of downward causation, Monica Anderson uses the example of a single word in our language, such as "like," standing alone in the middle of a blank page. What does it mean? The word "like" has about a dozen major meanings, so we need a context to determine its specific meaning. Then she goes on, "As you are reading a page, the words you are reading build up a high level context in your mind that influences how you interpret each individual word (low-level observation) that follows. This high level context exerts downward causation on the lower level word disambiguation process."

In a similar way, one could claim that a single musical note does not have a melody (or only an extremely trivial melody), but when you combine a series of musical notes together to create a song, that song can have a melody. As the experimental psychologist Irvin Rock rightly stated, "One can alter all the tones by transposing the melody in octave or key and still preserve the melody." A more realistic example of downward causation in science would be the following: A single nucleotide in DNA does determine the assembling of a protein, but its outcome is also determined by the surrounding nucleotides that somehow exert a downward causation on protein synthesis.

To take this discussion to an even higher organizational level: When people cast their votes in an election, it is not only the movements of physical particles in their fingers that determine the outcome, but also processes in their brains as well as very complex relations to their social environment. In other words, the emergent effects that cannot be observed in the individual components, such as finger movements, nevertheless affect the behavior of those components.

As to be expected, the idea of downward causation has been under attack as logically inconsistent. Jaegwon Kim says, "higher-level properties arise out of lower-level conditions, and without the presence of the latter in suitable configurations, the former could not even be there. So how could these higher-level properties causally influence and alter the conditions from which they arise?" This is not an issue that can be solved in this introductory book. But a possible, partial answer would be that we are not only dealing here with properties, but also with relationships between elements and properties, which are as real, and therefore causally effective, as those properties. Obviously, there cannot be any relationships if the elements on the lower level are missing. The components affect the network (upward), and the network affects the components (downward). So the interactions between lower-level constituents cannot easily be ignored. Because of this, similar causes can produce multiple effects, and similar effects may be the result of very different causes.

If there is indeed such a thing as downward causation, "wholes" would be able to enforce constraints on their "parts" to make them move in ways that may be unpredictable, even if there is complete information about the parts (ultimately the atoms and molecules) along with the complete information about the state of the universe outside those parts. Emergent properties of a higher organizational level seem to indeed cause the movements of their constituents. The simplest example of downward causation would be the move of an amoeba, causing all its constituent molecules to change their environmental positions—whereas none of them are themselves capable of such autonomous trajectories.

In cases like this, there seems to be a two-way causal flow between levels of organization. The systems biology approach, inspired by holism—and its related concepts of organicism, emergentism, and interactionism—avoids locating causation within a single, hierarchical level, and advocates a flow of causation that is bi-directional or reciprocal in nature. This is one of the reasons why higher-level processes—whether it is in the life sciences or the social sciences—are so difficult to understand. To paraphrase A. N. Whitehead, the various sciences study organisms of various sizes; sociology and biology would simply be the study of larger organisms, whereas physics studies smaller organisms.

In order to study certain phenomena, scientists use models, as we found out earlier (see chapter 1). The level they are working on and the direction in which they want to go—upward or downward—determine which kind of model they are going to use. Let me illustrate this with an example from cancer research that James Marcum uses. The bottom-up approach uses a somatic mutation model, in which researchers consider mutated genes at a lower level to be responsible for causing cancer. The top-down approach, on the other hand, uses a tissue organization field model to explain the cancerous disease. In the latter view, carcinogenesis is the result not of lower-order perturbations, such as mutated genes, but of perturbations of higher-order entities and their properties, such as disrupted or disorganized tissues or organs. Consequently, mutated genes would be the result, rather than the cause, of the disease's disrupted normal tissue architecture.

Either approach on its own is still one-directional. A two-directional approach would explain cancer in terms of a heterogeneous complex network composed of bidirectional or reciprocal interactions (bottom-up and top-down causation combined) among various hierarchical levels including genes, cells, and tissue architecture, organism, and environment. This requires what Edgar Morin calls a "shifting the perspective from the whole to the parts and back again." Traditionally, scientists tend to go for one-directional causation, and preferably of the upward type. Although many scientists have been trained to go the route of upward causation, the tide seems to be turning—which is the reason for my writing this book. The basic question in this debate this: Should a specific method determine the difference between good and bad science, or should the way science is actually done best determine what qualifies as a good method?

Fulvio Mazzocchi may have it right, when he says about the reductionist approach, "biologists might be reaching the limits of this approach. Despite their best efforts, scientists are far from winning the war on cancer, owing largely to the complex nature of both the disease and the human organism."

For further reading:

Anderson, Monica. "Reduction Considered Harmful." *H+ Magazine*   (2011), http://hplusmagazine.com/2011/03/31/reduction-considered-harmful.

Campbell, D.T. "'Downward Causation' in Hierarchically Organized Biological Systems." Pp. 179-186 in *Studies in the Philosophy of Biology*, ed. F.J. Ayala & T. Dobzhansky. Macmillan Press, 1974.

Marcum, James A. "Metaphysical Presuppositions and Scientific Practices: Reductionism and Organicism in Cancer Research," *International Studies in the Philosophy of Science* (2005), 19, pp. 31-45.

Marcum, James A., and G. M. N. Verschuuren. "Hemostatic Regulation and Whitehead's Philosophy of Organism." *Acta Biotheoretica*, 35 (1986): 123–133.

Von Bertalanffy, Ludwig. *General System Theory: Foundations, Development, Applications.* Penguin University Books, 1969.

II. Three Types of Reduction

The basic question of reduction is whether the properties, concepts, explanations, or laws from one scientific domain—usually at higher levels of organization—can be deduced from or explained by the properties, concepts, explanations, or laws from another domain of science, typically at a lower level of organization. So reduction is a one-directional approach that aims at bottom-up (upward) causation. Next, *reductionism* goes one step further than *reduction*, and claims that the technique of reduction is the only legitimate approach in science, thus in fact denying the possibility of any top-down (downward) causation. Whereas a model can be a simplifying tool at *any* level of organization (see chapter 1), reductionism simplifies the model further and limits it to a *lower*, or sometimes even the *lowest*, level of organization.

Would such a viewpoint be the end of *emergent* phenomena? Not necessarily so. Reduction and reductionism do not preclude the existence of what are called emergent phenomena, but they take it that those phenomena can be completely understood and explained in terms of the processes from which they are composed. However, this reductionist understanding of emergence is very different from what the term usually stands for—namely, that what emerges is more than the sum of the processes from which it emerges.

There are at least three different versions of reduction and reductionism: (1) a methodological version, (2) a theoretical (or epistemic) version, and (3) an ontological version. Let us see how these versions differ from each other.

4. Methodological Reduction

Methodological reduction is the most common form of reduction. It represents a technique of analyzing a complex system, located at a higher level, in terms of its simple components at a lower level. It promotes dissection and segmentation. Its "twin," methodological reductionism, adds to this that complex systems can only or should only be studied in terms of their components; it is not just selective but actually restrictive. Methodological reductionism is the position that the best, or even only legitimate, scientific strategy is attempting to reduce explanations to the smallest possible entities. Methodological reductionism would thus hold that the atomic explanation of a substance's boiling point is preferable to a chemical explanation, and that an explanation based on even smaller particles (quarks and leptons, perhaps) would even be better—if not ultimately mandatory.

Methodological reduction and reductionism explain things by analyzing them into smaller and simpler parts. It is a technique that has undoubtedly yielded a rich harvest of discoveries about the natural and social world. "Divide and conquer" is a basic scientific strategy for attacking any difficult problem: Take a large problem and divide it into smaller problems, then conquer the smaller problems one by one. As a means of analysis, then, reduction has certainly proven its value. It is natural to try to take any successful intellectual method as far as it will go. The technique of reduction allows a microbiologist to explain that a bacterium fails to respond to therapy because it has acquired a gene encoding a beta-lactamase enzyme, or that a patient exhibits enhanced susceptibility to infection because he or she has a mutant receptor for gamma interferon. This approach is what most people have come to expect from science.

The general idea of reduction is creating simple models describing smaller and smaller fragments of reality. If the extracted sub-problem is still too large to analyze, then we can repeat the process, dividing the extracted system farther down into smaller subsystems. If the scientific discipline we are operating in does not provide answers, then we might want to analyze the system in the "next lower discipline"—for instance, problems in sociology and psychology might be analyzed in terms of biology, and biological problems might be analyzed in terms of biochemistry, and so on. According to Sean Carroll, "The claim of reductionism is, depending on who [sic] you talk to, that the lower-level description is either 'always more complete,' or 'capable of deriving the higher-level descriptions,' or 'the right way to think about things.'"

Reductionism is a glorified version of reduction and fragmentation. The idea behind methodological reductionism is that complex systems should be investigated at the lowest possible level, and that experimental studies should ultimately be aimed at uncovering molecular and biochemical causes. Not only is this the most fruitful approach, but also the only legitimate one. It is undeniably a very common procedure of doing research nowadays: Psychologists have become neuroscientists who study human behavior at the neuronal level; physicians have turned themselves into cell biologists who study organisms at the cellular level; biologists have come to be molecular biologists studying cells at the molecular level. If you are looking for a research project, it is probably safer, smarter, and potentially more successful to go for a reductionist approach than a holistic one. But that is a rather pragmatic point of view.

Whether we honor the reductionist approach or not, it has to be granted that its technique of methodological reduction has been very successful in the life sciences, the behavioral sciences, and the social sciences. It allows their scientists to reduce a complex phenomenon—and almost all life phenomena and social phenomena are complex—to its simpler components, and this is usually the easiest way to gain access to them. Reductionists try to explain things by analyzing them into smaller and smaller parts—which is the "piecemeal" approach scientists are known to excel in. The reductionists' creed says: After dissecting them into smaller pieces, things are much simpler than they appear.

One word of caution, though. Do not mistake "simpler" for "easier to understand"; it is very doubtful whether an atom has a simpler "nature" than a molecule; smaller parts are not always really simpler, as quantum physics has shown us. Whereas Kepler's laws can be explained in a few minutes to a junior-high-school student, Newton's laws cannot be fully explained without using calculus. And to explain Einstein's theory requires four-dimensional, curved, non-Euclidean space-time, and much else besides. And yet, once we know more about it, Einstein's theory does have a compelling simplicity greater than Newton's theory. The simplicity to which scientific reduction leads us, then, is of a very paradoxical kind. It is a simplicity that is by no means simple-minded, as the particle physicist Stephen Barr puts it.

All of this is a powerful plea for using the technique of reduction. However, the power of methodological reduction that we have discussed so far may also obscure its weaknesses. One could arguably claim that exclusively reductionist research strategies can be systematically biased so as to overlook salient psychological or biological features— emergent features, if you will—and that for certain questions a more fruitful methodology consists in integrating the discovery of molecular causes with the investigation of higher level features. Interestingly enough, even though reductionists tend to explain organisms by their parts, they constantly, but incoherently, speak of organisms and thus keep appealing to organisms as a whole.

In other words, dissection of a complex system into its simpler parts may leave an unresolved residue behind. As A. Sengupta puts it, "Complex systems cannot dismantle into their parts without destroying themselves." Descartes may have thought that one can take the pieces of a whole apart, studying them, and then putting them back together to see the larger picture again, but is it really the larger picture again? Perhaps this method may work for machines, but what about living organisms?

The best known classical examination of the limitations of reduction is Erwin Schrödinger's book *What is Life*, in which he states that life and living phenomena cannot be reduced to the simple principles that physics has identified without losing the essence of life itself. This idea has been rephrased many times. Some would say, "If you try and take a cat apart to see how it works, the first thing you have on your hands is a non-working cat." Monica Anderson likes to say that "as the

Reductionist is cutting apart a living frog to 'see what makes it tick,' the 'ticking' disappears, but all the pieces of the frog are still there." Or take a rabbit out of its habitat into a laboratory and it will behave very differently. Or more in general, consider the laws of thermodynamics: As Monica Anderson simply states, "Any attempt to use them in the open world will fail because the energy interactions with the environment cannot be fully tracked."

That is why some would maintain that methodological reductionism is wrong when it claims that molecular biology is the only legitimate branch of the life sciences, or that neuroscience is the only legitimate branch of psychology. Adherents of the holistic, synthetic approach would claim instead that a scientist devoted to the reductionist, analytic approach is someone who, in the words of the late Nobel Laureate and biologist Konrad Lorenz, "knows more and more about less and less and finally knows everything about nothing." Besides, reduction is not the only way science is actually done. There is an increasing number of scientific studies not limited to a reductionist approach (see chapters 7 to 16). The core question here is: Should a certain technique such as reduction dictate what is considered "good science," or should good scientific techniques be determined by the nature and level of what a certain scientist is trying to study and achieve?

Take the example of epidemiology. It is a field certainly related to molecular biology, which in turn is related to chemistry and ultimately to physics, but—as Ferric C. Fang and Arturo Casadevall remark—the investigation of an ongoing cholera epidemic cannot be exclusively and effectively carried out at the level of a molecule of cholera toxin or the quantum state of an electron around a single carbon atom within the toxin B subunit. That is hard to argue with. Something similar holds for a discipline such as immunology. Limiting oneself to immune cells in test tubes (*in vitro*) gives us the false impression that the immunity system is really an isolated, self-regulating system that operates without other bodily systems (*in vivo*).

There are times that the best way to study the behavior of a complex system is to treat it as a whole, and not merely to analyze the behavior of its component parts. In such situations, an understanding of a certain kind of complex system is best sought at the level of principles governing the behavior of the whole system, and not at the level of the structure and behavior of its component parts. Those who forget or refuse to do so may have left behind some context that would have affected their results. This could be called a "reduction error." Such errors occur, for instance, when choosing a model that ignores friction when it actually matters, or when ignoring any emergent effects in complex systems. Or, to use a simpler example, when you take an analog watch apart, you have no watch left, but only a series of pieces that do not do anything until you are able to put them back together in the proper way. If you are unable to do so, you have a rather dramatic "reduction error" at your hands.

Next let us address the question of why reduction is so popular in science and has been turned by some into a dogma of reductionism. Why is the reductionist approach so common in science, whereas the holistic approach still remains more the exception? There are probably several reasons.

First of all, we have a success story here. Reduction is definitely a useful and effective method in science, because it breaks the object under investigation down into a set of smaller and smaller components which can be easier studied in isolation than within the complex setting of the original. Although models can be used on all levels of organization and complexity, it is often easier to construct and manipulate models at the lower levels. It must be stated in all fairness, though, that those who use reduction do not necessarily deny complexity, but they are looking for and focusing on the simple rules that underpin it. No wonder then, the goal of any science and engineering education nowadays is to give the student the ability to "perform reduction." It is a training in "piecemeal approach." It has become the most commonly used process in science and engineering, and we silently assume we will and should use it at every opportunity we have. Therefore, there has been little need to discuss reduction as a topic outside of epistemology and philosophy of science.

Second, aspiring scientists are being brought up with a particular paradigm. The term paradigm, which Thomas Kuhn introduced, has been defined in many ways, but it typically stands for a collection of rules on how to solve scientific puzzles. Most scientists feel attached to the paradigm they were brought up with. The reason is that individual scientists acquire knowledge of a paradigm through their scientific education—that is how they learn their standards, by solving "standard" problems, performing "standard" experiments, and eventually by doing research under a supervisor already skilled within the paradigm. Aspiring scientists become gradually acquainted with the methods, the techniques, and the presuppositions of that particular paradigm. And the technique of reduction is typically part of that paradigm. Because of their training, scientists are usually unable to articulate the precise nature of the paradigm in which they work, until a need arises to become aware of the general laws, simplified models, metaphysical assumptions, and methodological principles involved in their paradigm, and forces them to think "outside the box."

Third, there is also some kind of "logic" behind the idea that reduction and reductionism should be promoted in science. One of them is a particular, rather narrow conception of causality, according to which each event must have a sufficient *physical* cause. Physical interactions are therefore considered sufficient to account for all causal interactions. Under this assumption, psychological or biological facts should be replaced by physical facts, given that the physical conditions are supposed to do all the causal work. This idea tends to exclude any other, non-physical conditions as causal factors.

Another "logical" defense of using reduction and reductionism comes from the principle of "Occam's razor." This principle states that among competing hypotheses, the one with the fewest assumptions should be selected. Other, more complicated solutions may ultimately prove correct, but—in the absence of certainty—the guiding rule should be: the fewer assumptions that are made the more accurate the analysis is likely to be. However, a word of caution is needed here. Although Occam's razor is often taken as a plea for reduction and reductionism, at times it may also be taken as a defense for holism. Sometimes the "simplest" explanation is not to be found on lower levels but on higher levels. Take, for instance, orientation behavior of homing pigeons. Is it really "simpler" to explain this behavior in terms of stimulus-response actions, or would it be "simpler" to speak of cognitive maps that the birds use? The answer to this question probably depends on more than Occam's razor.

In other words, simplicity may work in opposite directions: A simpler description may refer to a more complex hypothesis, and a more complex description may refer to a simpler hypothesis. As a consequence, so Peter S. Williams warns us, full-fledged reductionism "is a bad habit because the legitimate search for the simplest unifying explanation is pursued with such dogmatic commitment that the demand for explanatory simplicity outweighs the primary requirement that explanations must be adequate to the nature of the data they are meant to explain. The facts are reduced to fit a single, simplistic, inadequate explanation, rather than explanation being expanded or multiplied to fit the facts."

To sum up this discussion, it is hard to discredit reduction as a technique, for it has a great track record. However, it is also hard to claim that reduction is a complete, comprehensive, all-embracing method that leaves no room for any additional methods. Reduction may in fact not do enough justice to the complexity of emergent properties at higher organizational levels. Hence there may still be a need for top-down causation in addition to bottom-up causation. To use a simple analogy, that which is happening on a computer screen can be explained "bottom-up" by the forces of all electrons involved, but also "top-down" in terms of software. As a popular saying goes, "As long as you are looking down, you cannot see something that is above you."

Let us come to a conclusion by quoting what one of the Linus Pauling collaborators, chemist Arthur Robinson, said: "Using physics and chemistry to model biology is like using Lego blocks to model the World Trade Center." The instrument is simply too crude. We need science without dogma.

For further reading:

Ayala, F.J. 2011. Philosophical Issues in Biology: Reduction and Emergence. Pp. 137-161 in *Philosophy in Science.*

Methods and Applications, ed. B. Brozek, J. Maczka, and W.P. Grygiel. Kraków: Copernícus Center Press, 2011.

Jones, Richard H. *Analysis & the Fullness of Reality: An Introduction to Reductionism & Emerge*nce. Jackson Square Books, 2013.

Kirschner, M.W. "The meaning of systems biology." *Cell* 121:503–504, 2005.

5. Theoretical Reduction

Theoretical reduction is a more sophisticated form of reduction. It is—at least in outset—a technique of deriving a higher-level *theory* from a lower-level *theory*. Theoretical reduction is just a technique in the formation of theories, but it may be grounded in the conviction that theoretical reduction is the only way to understand and explain nature in theoretical terms. This latter conviction would amount to theoretical reductionism—being a more dogmatic version of theoretical reduction.

The program of theoretical reductionism is based on the belief that there is a unity in all natural phenomena; this unity is supposed to show itself in the unity of fundamental concepts, laws, and theories. So the claim is that all higher-level theories should be derived from low-level theories in order for them to be scientifically acceptable. Ultimately, this form of reduction entails that all scientific theories either can or should be reduced to a single *super-theory* through the process of theoretical reduction. The general idea is that descriptions in terms of entities that are literally smaller than the others— atoms are smaller than neurons, which are smaller than organisms—tend to have a wider range of validity.

We are dealing here with a rather technical, high-brow enterprise that was initially applied to theories only. It led to the position that all scientific theories either can or should be reduced to a single super-theory through the process of theoretical reduction. Ultimately, this view tends to come down to the assumption, or at times even a conviction, that the laws of fundamental physics are universal in scope. Because of this, any laws claimed to apply to complex entities would have to be deducible in principle from these fundamental laws—or be false. Once the aim of reducing theories concerning a higher level to theories concerning a lower level has been achieved, the axioms of the reduced theory have become *theorems* of the reducing theory.

In other words, theoretical reduction is the process by which one theory absorbs another. For example, both Kepler's laws of the motion of the planets and Galileo's theories of motion as used for terrestrial objects are reducible to Newtonian theories of mechanics, because all the explanatory power of the former are contained within the latter. Furthermore, the reduction is considered to be beneficial because Newtonian mechanics is a more general theory—that is, it explains more phenomena than Galileo's or Kepler's. Then in turn, Newton's theory can be subsumed under, or reduced to, Einstein's theory. Something similar can be said about deriving Boyle's law from the kinetic theory of gases, although some have pointed out that this only holds for the van der Waals equation, so not Boyle's law itself, since the former is slightly more accurate than the latter. Yet, one theory can be seen as "absorbing" the other.

Something similar may perhaps be achieved in genetics. Since Gregor Mendel, there is a genetic theory on the level of organisms (TO), dealing with visible, phenotypical traits. Later on, Thomas Hunt Morgan framed a theory on the level of cells (TC), which provides an explanation for Mendel's organismal theory in terms of genes located in chromosomes inside cells. And recently, we acquired a genetic theory on the level of molecules (TM), which allows for rendering genes in terms of DNA-fragments. If it were possible to link TC-statements about genes to TM-statements about DNA-fragments, TC could be considered to have been "reduced" to TM. The ultimate aim of theoretical reductionism is to find a physico-chemical terminology that could fully replace the biological terminology of genetics.

However, reducing higher-level theories to lower-level theories has become increasingly difficult, and even questionable. Some have argued that the relation between classical and molecular genetics, for instance, is one of theory replacement, rather than reduction in the strict sense. Others have argued that molecular biology as the reducing theory does not consist of a small body of laws, so that a covering-law model of explanation and reduction is not applicable here. Besides, it is doubtful whether a reduction of theories is a good fit for the way biology has actually been developed and practiced. For example, the advent of the chromosome theory of inheritance in the 1910s bridged the previously unrelated fields of Mendelian genetics and cell biology. This "inter-field" theory did enable a unification of these two fields, but Mendelian genetics and cytology were not reduced to each other, nor did the inter-field theory reduce both fields. This seems to make theoretical reductionism a rather questionable goal.

The following analogy might help explain more why some believe reduction of theories is problematic. The laws of syntax, governing the grammatical construction of sentences, allow for many arrangements of words, but the laws of semantics govern which subset of these arrangements actually has a semantic meaning, and the laws of rhetoric in turn govern which sub-subset is good oratory. It follows that the laws of rhetoric cannot be reduced to the laws of semantics, and these in turn cannot be reduced to the laws of syntax. Would it not be absurd to claim that an understanding of Shakespeare's works can only be achieved by studying the English alphabet? Something similar may hold for the laws of the life sciences and the social sciences in connection with the laws of physics.

In addition, the idea of a "super-theory" that can absorb all other theories has become under attack. It can in fact be argued that in practice disciplines such as physics, biology, psychology, and sociology are epistemologically discontinuous, for as Ferric C. Fang and Arturo Casadevall put it, science currently lacks a grand theory that allows us to connect such disparate phenomena as quantum mechanical states and the songs of birds and the thoughts of humans and the culture of a society. Yet, it remains a timeless temptation to search for a "Grand Theory of Everything." Is that not something physicists are dreaming of—the "grand unification" or "grand unified theory" (GUT)?

Admittedly, there is a strong argument in favor of the universality of physical laws. The reasoning goes along these lines: All material systems are governed by the laws of physics; all living systems are material; therefore, all living systems are governed by the laws of physics. Just as the gas laws have been reduced to the laws of kinetic theory, so must all laws "in the end" be reduced to one fundamental level. This kind of reasoning seems to force us into theoretical reductionism. But perhaps not necessarily so. Obeying the laws of physics does not exclude obeying other laws as well. There is ample reason to study complex systems also at their macro-levels, with their own laws, and in their own terms, which are laws in addition to physical laws. Why forbid such investigations in the name of an abstract idea—"the unity of science"?

Here is another word of caution. Physicists are merely in search of an "umbrella" theory in physics that would unify the three non-gravitational forces as aspects of one single underlying force. No matter what the outcome will be, a search for a physical "umbrella" theory may be a very legitimate undertaking, but such a "grand unified theory of *physics*" is not the same as a "grand theory of *everything*." (A theory of *everything* would also have to explain why some people believe in it and some do not.) Even if such an "umbrella" theory would ever be found, it would explain only all the facts *physics* deals with in this universe, but it would never amount to a super-theory fully explaining everything else in this world as well. In other words, it would not qualify as a "Great Theory of Everything in this Universe," so to speak, unless we assume already that everything has to be reduced to physics—which is exactly the issue under debate.

The reason for being cautious about complete unification is rather simple. Science works with models, and reduction works with models. However, models always provide *partial* representations of aspects of reality that provide only *partial* insight into the complexities of the real world; they are never replicas of the original (see chapter 1). Using the model of a computer, for instance, to study the working of the brain as if it were a computer does not make the brain a computer. The idea that laws of physics explain all aspects of reality is essentially a reductionist viewpoint. According to Marjorie Grene, on grounds of both method and content, science is partial and plural—it is "irreducibly pluralistic." So the pursuit of a "grand unification of everything" is most likely an illusion.

Of course, one could object that we have no right to say what physics will *never* achieve, because that would be an a priori answer to an empirical question. However, those who consider that question empirical are plainly arguing a priori themselves. What we could say, though, is that because anything may be relevant to anything, science is "one," but at the same time also "many." That would make for unity in plurality.

For all the above reasons, the goal of theoretical reduction has been revised by some. Instead of reducing higher-level *theories* to lower-level theories, the less ambitious aim has become reducing higher-level *explanations* to lower-level explanations. This is sometimes more carefully rephrased as epistemic or explanatory reduction. In its ideological form, it asserts that all explanation must be one-leveled in terms of least particulars—whatever these may be.

In this way, according to Ingo Brigandt and Alan Love, many mechanisms of classical genetics can be reductively explained by the physical principles governing the behavior of macromolecules (though approximations are still necessary). Explanatorily relevant principles for the physics of macromolecules include weak interactions, hydrogen bonding, hydrophobic bonding, allosteric transformations, lock-key fit, and the idea that structure determines function. By using the molecular mechanisms of replication, recombination, and cell division, we might be able to provide reductive explanations of the principles of classical genetics. As Brigandt and Love put it, "While classical genetics offered an account of gene transmission and could also rely on cytological explanations, these accounts did not yield a complete explanation of how and why recombination occurs. Molecular biology filled in this gap and corrected classical genetics."

Although the new goal is less ambitious than before, it still seems to face several potential problems. A first problem would be that there is no one-to-one relationship between, for instance, a Mendelian gene and a molecular gene. The early molecular gene did have a beginning and an end, but it turned out to be a stripped gene, namely a structural gene (also called an expressive gene) as part of a wider architectural setting that can vary greatly. Further research has revealed that structural regions can be shared, overlapping, nested, and even physically split; moreover, alternative slicing can produce multiple products of gene translation, plus one product can have multiple functions. Thus, a molecular gene is not a well-defined structure. A molecular gene can be divided up into several domains (such as enhancer, promoter, suppressor, intron, exon, and the like), but none of these domains is a necessary part of a gene. In short, what makes for a unit in Mendelian genetics is not necessarily a unit in molecular genetics; rather, a molecular gene is a collection of component domains. In other words, the units of molecular genetics do not nicely match the units of higher biological levels on a one-to-one basis (see chapter 10). There isn't even a one-to-one relationship between the genes and the proteins they codify, as the number of proteins exceeds the number of genes.

The same holds for other biological phenomena. For example, there are several physico-chemical ways of carrying out some particular biological function, such as oxygen transportation in the blood. Most animals have a red oxygen-carrying pigment called hemoglobin, whereas many mollusks and arthropods have a blue pigment called hemocyanin, which contains copper instead of iron. In the course of evolution, different lineages of organisms have come to solve similar functional problems in very different physico-chemical ways. This means that biological properties do depend on physico-chemical properties, but not on the basis of a one-to-one relationship.

Reversibly, there are only some out of many physico-chemical arrangements which perform a biological function. The twenty different building blocks of proteins, for instance, can be put together in many different ways, but only certain arrangements of amino acids have a biological function. There is definitely a physico-chemical explanation for the fact that these biological structures actually exist and work the way they do, but— because physico-chemical laws allow for more structures than just the functional ones from biology—there is really no physico-chemical explanation for the way biological structures came into existence and attained functionality. Additional explanations are needed.

In addition to many-one relationships, there may also be *one-to-many* relations between molecular kinds and higher-level kinds. A molecular pathway, for instance, may have different effects in different cellular contexts, so that the same pathway can be involved in different functions in different species or in different parts of an organism. Even the amino acid sequence produced by a molecular gene may depend on DNA elements outside of this gene and on non-genetic factors, so that an expressive gene can code for distinct products in different cells or in different states of a cell. Thus there is a one-to-many relation between molecular kinds and higher level kinds, since a molecular mechanism can causally lead to or be part of different higher level states depending on the context. One may counter-argue, of course, that one could specify the relevant context as initial conditions in the molecular premises from which the higher level state is to be deduced.

In the case against theoretical reductionism, there is one more argument that needs to be addressed. We discussed earlier that many concepts are specific to a particular organizational level. For example, describing a concept like "insulin" in terms of a molecular structure may be a good representation of the denotation of insulin, but it does not do justice to the connotation of the concept. Two concepts with the same denotation may have very different connotations; "morning star" and "evening star," for example, have the same denotation, namely Venus, but different connotations. Thus, a biological concept such as "hormone" may be related to a chemical object (its denotation) which can be produced in a test tube, but at the same time it has a biological meaning (its connotation) relating to the specific activity this molecule exerts within an organism. Without organisms, no molecule would qualify to be called a hormone. And the same holds for processes at higher levels such as predation, frustration, and enculturation; all of these are processes specific for a certain organizational level. All of this would raise even more questions if we were to extend the discussion into the domain of the behavioral and social sciences.

Therefore, we might come to the following conclusion. Explanations and concepts relating to lower levels may be important or even necessary in order to understand processes on higher levels, but they may arguably not be sufficient. The laws of physics, for instance, must permit behavior on higher levels—for a higher level cannot violate the laws of lower levels. However, it is also true that lower levels have to go along with the actions on the higher level. In other words, biological structures are definitely determined by physico-chemical laws, but they are under-determined, that is to say, not uniquely determined. Even if some arrangements are chemically more likely than others, the persistence of one particular arrangement, instead of any other, is not explicable in chemical terms alone. Out of a large number of possible arrangements, all of which are equally compatible with the laws of physics and chemistry, only a few arrangements have a biological, psychological, or sociological function; some perform one and the same function, others do not perform any such function at all. If this is correct, then there is no simple one-to-one relationship between physico-chemical properties and properties at higher levels of organization.

Besides, the lowest-level explanation of a phenomenon, even if it exists, may not always provide the best way to understand or explain the higher-level phenomenon. The lower level leaves open "boundary conditions" to be specified by the laws of the higher level. Marjorie Grene summarizes this as follows: "The laws of the lower level cannot explain the existence of the higher level—although they may suffice to explain its failures."

Finally, we need to realize that the zeal of theoretical reductionism raises practical problems. In order to reduce ecological theories about predation to physico-chemical theories, the concept of predation has to be reduced to physico-chemical concepts first. However, the concept of predation is related to an enormous variety of phenomena—such as the consumption of a gamut of food (from grass to insects) by a gamut of organisms (from bacteria to elephants). For research purposes, a physico-chemical "definition" including all of this would hardly be manageable. Concepts and classifications which are appropriate at a molecular level may be extremely clumsy at an ecological or psychological or societal level. That would make theoretical reductionism not even worth our time.

For further reading:

Anderson, P.W. "More is Different." *Science* 1972, 177 (4047): 393–396.

Brigandt, Ingo and Love, Alan, "Reductionism in Biology", *The Stanford Encyclopedia of Philosophy* (Winter 2014 Edition), Edward N. Zalta (ed.). Here is the, URL link: <http://plato.stanford.edu/archives/win2014/entries/reduction-biology/>.

Schaffner, K. F. "Approaches to reduction", *Philosophy of Science,* 1967, 34:137–147.

Simon, Michael A. *The Matter of Life: Philosophical Problems of Biology*. New Haven, CT: Yale University Press, 1971.

6. Ontological Reduction

Ontological reduction is the idea that reality is composed of a minimum number of kinds of entities or substances. Whereas ontological reduction is a mere assumption, ontological reductionism is based on the conviction that reality is ultimately nothing more than a minimum number of basic substances—typically atoms and molecules. It is a form of atomism—based on a creed that says, "atoms are not only everywhere, but they are also all there *is*." It is a claim about what reality ultimately is composed of.

Ontological reduction reduces the number of ontological entities to a very small set of "primitives" that exist within an ontology. This supposedly simplifies philosophy, because every ontological primitive requires its own description and explanation. To define or explain life, for instance, would have to be done in terms of what is not alive. But if we declare life as a real property, then we must give a separate explanation of why some objects possess it and why others do not; this sort of process can prove to be quite complex. Additionally, one would have to prove that a primitive is actually worthy of this status, instead of being defined as a variant of something else which is more basic.

So the idea behind ontological reduction is to unify and simplify our ontology, while guarding against needless multiplication of entities in the process. Therefore, ontological reductionists usually end up with a one-level ontology, making all the higher levels either illusionary or arbitrary, usually by demoting them to a mere scheme of classification. To say that certain entities are "alive" is believed to indicate merely that they exhibit qualities that distinguish them from other entities we would not describe as being alive. But sometimes it is true that the question of what it means to say that an organism is alive can best be answered by abstracting from the fact that it is alive.

The claim of ontological reductionism is of a metaphysical nature, in effect claiming that all objects, properties, and events are reducible to a single type of substance. In metaphysics, this idea is often called physicalism, or materialism, which assumes that there is no difference in higher-level properties without a difference in some underlying physical properties—or put in metaphysical terms, that each particular higher-level process is metaphysically identical to some particular physico-chemical process. This poses the question, of course, as to whether there is still room left for genuinely emergent properties—even more so because ontological reductionism usually denies that interactions between physical entities are real and can in fact create emergence.

Obviously, ontological reduction would involve the question as to what the "real" entities are in scientific analysis. What about substances, what about properties, and what about relationships? Some reductionists accept only substances and their properties as being real, but deny that relationships are also "real." Interactionism, on the other hand, assumes that relationships are "real" as well. Obviously, stands like these are essentially of an ontological, metaphysical nature.

Ontological reductionism in its most common form tends to make metaphysical, ontological *is*-statements in the form of "X *is* nothing but matter in motion." It is the source of slogans like "Life is nothing but chemistry and physics" (James Watson); "Humans are nothing but a speck of dust" (Carl Sagan); "Humans are nothing but a pack of neurons" (Francis Crick); "The brain is nothing but a 'meat machine.'" (Marvin Minsky); "Humans are nothing but glorified animals" (Charles Darwin); or "Humans are nothing but a bundle of instincts" (Sigmund Freud). In all these cases, the words "is" and "are" feature as keywords. However, one could argue against claims like these that within the setting of a biological model, human beings may indeed be seen as "*only* DNA," but in reality they are "*also* DNA." This argument assumes that ontological reduction may be fruitful for research purposes, but it does not automatically lead to ontological reductionism.

According to ontological reductionists, all societies are "in reality" just a collection of human organisms, all organisms are basically only a collection of cells, all cells are merely a collection of molecules, and ultimately all molecules are just a collection of atoms. As a consequence, an ontological reductionist would see psychological facts as fully reducible to neurological facts, which in turn are fully reducible to biological facts, and ultimately to facts about sub-atomic particles. In this view, psychology should therefore be explained in terms of biology, biology should then be explained in terms of chemistry, and chemistry should ultimately be explained in terms of physics. This reductionist view declares the apparently heterogeneous structure of the universe to be basically and essentially homogenous. It changes psychology into applied biology, biology into applied chemistry, and chemistry into applied physics.

This tendency to downgrade and diminish reality is considered by someone like Stephen M. Barr as "a metaphysical prejudice that equates reduction with a grim slide down the ladder of being." Ontological reductionism assumes that things *are* actually simpler than they appear. However, what simpler means in science is much discussed among philosophers—it is not at all a simple question. But to many materialists it seems to mean lower, cruder, and more primitive. By this way of thinking, the further we push toward a more basic understanding of things, the more we are immersed in pieces of *matter*. This makes some claim that any ultimate explanation of mental phenomena will have to be in material terms, or else it won't be an explanation at all. Of course, the logic of this could also be turned around. One could just as well say that any ultimate explanation of the material world must be in non-material terms. But materialists won't easily give up their conviction that the lower explains the higher.

Those who accept the doctrine of materialism—"Matter is all that matters"—adhere to an extreme form of ontological reductionism. But it could be argued that materialism gets the direction of reduction wrong: Sometimes only the "whole" can account for the properties of its parts, but not vice versa. Kristor Lawson uses the example of a salt molecule to make this case. A complete account of a salt molecule must include a full specification of the properties of its sodium atom; but a complete account of a sodium atom on its own cannot include a full specification of the properties of a salt molecule.

In contrast to materialism—with its one-level ontology—holism usually adheres to a *multi*-level ontology by acknowledging the real existence of entities and their emergent properties on any higher level. Even those who accept methodological reduction—or theoretical reduction, for that matter—are not forced to also accept ontological reduction on a one-level basis. Just because we *can* think of something macroscopic in terms of its microscopic parts (which is methodological reduction), that does not mean such a macroscopic thing becomes any less real (to claim otherwise would amount to ontological reduction). In a holistic view, cells and social structures exist in the same way as chairs and tables exist. They are real entities.

Holists who oppose ontological reduction would stress that excessive reduction can lead to oversimplification. For example, there is a clear distinction between animal life and plant life, as animals have powers that plants lack, such as sensation, active locomotion, and arguably emotionality. Reducing all of these to the same thing would then confuse our thinking and talking about animals or plants. Ontological reductionism can thus erase real and important distinctions.

Some metaphysicians and philosophers of mind would go even further. They contend that there are strong first-person, introspective grounds for supposing that consciousness, intentionality, cognitivity, and/or human agency are cases of ontological emergence. The intrinsic qualitative and intentional properties of human experiences, so they suggest, appear to be of a fundamentally distinct character from the properties described by the physical and biological sciences. (Whether this warrants a distinction between matter and spirit is another question; see chapter 13.)

To put it differently, ontological reductionism is most commonly rooted in *monism*, in effect claiming that all objects, properties, and events are reducible to a single substance— matter, that is. In general, monists are right in claiming a single fundamental category suits the unity of all sciences best—it makes for something all sciences have in common. However, theoretically, there could also be a version based on *dualism* claiming that everything is reducible to one of *two* kinds of entities—either matter or spirit. Needless to say that many scientists typically reject any kind of dualism. Instead, they opt for a one-level ontology, which is usually the level of physics, making mental phenomena a by-product of material phenomena. One should ask, though, if there is "really" only one kind of science; if so, it would in principle be physics. Is biology, for instance, "really" a molecular science—a science of mere aggregates of molecules, atoms, or sub-atomic particles? Or is there more to it?

Some would dispute the narrow ontological claim of materialistic monism. They argue that the claim of a one-level ontology would in fact contradict itself, because this very claim is a matter of cognitivity in itself. Hence, we seem to rather need at least a two-level ontology that acknowledges cognitive events in addition to molecular events in order for us to make any claims of a cognitive nature. Even if molecules are all there is to know, and if there is knowledge, there must be something more than physical particles in motion. Molecules can make no claim to truth, any more than they can err. If there is any knowledge, even in "molecular science," there must be something more than what molecular science is about. Either there is no knowledge, or there is at least the knowledge that one-level ontologies such as physicalism and materialism are false. So, if there is knowledge, then there must be at least more than one level of reality. Otherwise the claim of reductionism can only be uttered as a noise but not as a claim to real knowledge. If reductionism were true, so the reasoning goes, knowledge would be impossible, including the knowledge that reductionism is true.

If this is a valid objection, we would end up with an ontology of at least two levels—molecular events, on the one hand, and cognitive events, on the other hand. But there could be—and arguably is, according to holism—much more in between, which would call for a multi-level hierarchical structure in ontology (see chapter 13). But no matter which position we take, we need to admit that materialism and physicalism are "viewpoints" based on a certain philosophical position; they cannot be conclusions of empirical research. The sciences can corroborate that there is matter, but they cannot possibly affirm that there is nothing else but matter. Once you limit yourself to "matter," you cannot claim at the same time that there is nothing else but "matter."

For further reading:

Dupré, J. The Disorder of Things: Metaphysical Foundations of the Disunity of Science. Cambridge, MA: Harvard University Press, 1993.

Koestler, Arthur, and J. R. Smythies, ed. *Beyond Reductionism: New Perspectives in the Life Sciences*. London: Hutchinson, 1969.

Ruse, Michael. "Do Organisms Exist?" *Amer. Zool.*, 29: 1061–1066 (1989).

Silberstein, Michael, and John McGeever. "The Search for Ontological Emergence." *The Philosophical Quarterly*, 1999, Vol. 49, No. 195.

III.   Beyond Reductionism

It is safe to say that reduction, in any of its forms, has proven to be a powerful and successful method for scientific research. But it is also legitimate to question whether reductionism is right in claiming to be the only appropriate method for scientists. The choice for a reductionist or a holistic approach does not have to be a matter of "either-or" and will most likely not be solved by asking "Which is the only one to work?" The main reason is that both parties disagree over what is to be counted as evidence; a radical reductionist, for instance, will claim ahead of time that a holistic approach cannot possibly work.

One thing is clear, though: It has become more and more difficult to conceptualize a single, adequate conception of reduction that will do justice to the diversity of phenomena and reasoning practices used in the life sciences, the behavioral sciences, and the social sciences—as we will see in the chapters to come. The multiplicity and heterogeneity of sub-disciplines we find under the wide umbrella of what is called "science" only reinforces this argument and suggests to some that we should move beyond reductionism entirely. A one-sided approach arguably would constrain rather than liberate the sciences.

What probably, and hopefully, became clear in the previous chapters is that there is no simple pro-or-con position regarding the reductionist approach versus the holistic approach, or on the subject of reductionism versus holism. The difference between the two is like choosing between a microscopic view and a macroscopic view—put in an image, to either see the trees or see the forest for the trees. Either one of the two has its advantages and disadvantages, its capabilities and its weaknesses. Some favor a "nothing but" approach, some an "over and above" attitude; it is more or less like a quarrel of opinions over what is "more basic" and what is "more important."

Individual scientists may have a preference for one of the two, but an exclusive claim seems hard to defend. Downward causation does not, in and of itself, preclude upward causation, and upward causation does not preclude downward causation. It is even possible to study bi-directional "traffic" between two hierarchical levels (see chapter 3). All of this may plead for a two-directional approach, at least in principle, depending on the context of the specific system or phenomenon that is being studied. Again, it is a matter of "both-and" rather than "either-or." As Brigandt and Love summarize it, "Reductionist strategies in science attempt to break down systemic dynamics into different components, whereas holistic strategies aim to understand how the quantitative interactions among lower-level entities result in higher-level behavior."

A case in point would be the process of homeostasis in organisms. It channels within rather narrow margins accidental changes in the internal environment of an organism, such as changes in temperature, pressure, and acidity in the body. On the one hand, a phenomenon like this can certainly be analyzed at lower organizational levels, such as glucose levels and pH in the blood. On the other hand, homeostasis is also a candidate for a more holistic approach, because it requires higher-level norms regulated by built-in "thermostats" and feedback loops. It is hard to see how one approach could be sufficient without the other.

As we found out, the universe is stubbornly hierarchical (see chapter 2). George Gilder describes it as "a top-down 'nested hierarchy,' in which the higher levels command more degrees of freedom than the levels below them, which they use and constrain." Biological phenomena, for instance, do rest on the physico-chemical level; they are constrained by physico-chemical laws, but these, in turn, cannot fully explain what happens at a higher level. Although the sciences can be arranged roughly in a linear hierarchy—in that the elementary entities of one science obey the laws of the science that precedes it in the hierarchy—yet this does not automatically imply that one science is just an applied version of the science that precedes it. At each level, entirely new laws, concepts, and generalizations seem to be necessary, requiring inspiration and creativity to just as great a degree as in the lower level. As the Nobel Laureate Philip Anderson puts it, "Psychology is not applied biology nor is biology applied chemistry."

To paraphrase Ludwig von Bertalanffy: Growing up under the shadow of physics and chemistry, many sciences have languished like plants deprived of lights. This view questions what gives hard-core reductionists the right to claim that the behavioral sciences, for instance, can only become "genuine" scientific disciplines if they are based on molecular biology. Such an extreme claim may in fact have been made but is hard to justify.

To give any multi-level approach a more philosophical underpinning, let us examine how the late Hungarian-British physical-chemist and philosopher Michael Polanyi would deal with the separate levels of the organizational hierarchy. He uses the machine analogy. His thesis is that a machine based on the laws of physics is not explicable by the laws of physics alone, because the *structure* of a machine is what he calls a "boundary condition" extraneous to the process it delineates. Polanyi uses the term *boundary condition* in a very specific way. Boundary conditions are not understood here the way we used them earlier in relation to models (see chapter 1) or the way they relate to laws of physics and chemistry—as in the Ideal Gas Law, for instance, where extreme temperatures and pressures are considered boundary conditions.

Instead, when Polanyi uses the term, he means that the structure of a system acts as a boundary condition that "harnesses" the way it functions. Any kind of mechanism has a specific configuration that represents only one of a very large number of possible arrangements, all of which are compatible with the laws of physics and chemistry. In other words, even machines are not one-level but multi-level systems (see chapter 16). Besides, an account of how a system works is not the same thing as an account of how it came into existence.

Polanyi insists that this view can also be applied to the organizational hierarchy of nature. A higher level creates, so to speak, "boundary conditions" for the lower-level components. That is the reason why a lower-level entity such as DNA can properly function, because it is "harnessed" inside cells located at a higher level, and those cells, for their part, can function even more intricately when they are harnessed in organisms at a still higher level. Polanyi teaches us that principles and processes at a higher level cannot be derived from principles and processes at a lower level *alone*, just as it is impossible to derive a grammar from a vocabulary, or a vocabulary from phonetics alone.

What we have here is an organizational hierarchy in which the higher levels command more degrees of freedom than the levels below them, which they use and constrain. Thus, the higher levels can neither eclipse the lower levels nor can they be fully reduced to them. Obviously, all higher-level fields do in fact *depend* on chemical and physical processes, yet they are not uniquely *defined* by them. Operating farther up the hierarchy, macro-systems such as brains, minds, human beings, businesses, societies, and economies consist of higher entities that harness chemical and physical laws to higher-level purposes but they are not entirely reducible to lower-level entities or fully explicable by them. The lower level leaves open boundary conditions to be specified by the laws of the higher level. The laws of the lower level cannot explain the existence of the higher level—although they may suffice to explain its failures.

The key ingredient that endows a collection of parts with causal efficacy is *organization*—which is expressed in keywords such as organicism and interactionism. Components, whether cells or transistors, acquire new "causal power" when they are organized into a mechanism that performs a higher-level function. Earlier we used the example of two resistors in a circuit: Whether they are connected in series or in parallel makes a big difference for their total resistance. So the new causal power does not lie in the physics of the components alone, which typically does not change when the components are organized together; the secret lies in the organization per se which exploits the physics to accomplish a functional goal. Walter Sinnott-Armstrong formulates it this way: "[P]hysics constrains but does not determine function."

Interestingly enough, it was the German pathologist Rudolf Virchow who revolutionized the life sciences by comparing an organism with the "holistic" structure of a human society, thus making the cell an organism's fundamental building block that interacts with other cells in specific ways. By declaring that cells can only arise from other cells, he made them elements of a "society" of cells that act together in a "societal" way. In this view, an organism is like a "republic" of cells, a well-organized collection of tiny individuals making up a single super-individual. This opened the way for future concepts such as homeostasis combined with hormonal and neural integration.

Once we acknowledge the reality and existence of multiple levels, terms such a reductionism and holism become to some extent relative. In the words of ecologist Rick C. Looijen, a phenomenon, property, or theory may be "reductionist with respect to some higher level, but holistic with respect to some lower level." Therefore, the results of a holistic approach that was started at a specific level could also be reached with a reductionist approach when done from the next level up.

As a consequence, the difference between upward and downward causation is basically a matter of perspective within the framework of a hierarchy of life. What does remain standing, though, is that higher levels do reveal emergent properties that fail to surface at lower levels. In general, one might say that a reductionist approach and a holistic approach are not mutually exclusive but complementary instead. Either way seems to be legitimate and may open up avenues that its counterpart might easily miss on its own. Hence, we should leave room for explanations based on downward causation as well as explanations based on upward causation. This does not rule out, of course, that individual scientists may have their own strong preference for either one in particular.

There is a growing awareness of this methodological duality in the scientific community. More and more scientists are beginning to realize that the scientific investigation of biological, behavioral, and social phenomena demands a diversity of tools and methods without a requirement to tie it all together into fundamental physics, macromolecular chemistry, or molecular biology. While it is easily accepted that the foundations of chemistry are based in physics, and that molecular biology is rooted in chemistry, similar statements become rather controversial when one claims that sociology and economics are fully based on psychology, or that psychology is entirely based on neuroscience. Such claims seem to overstep their own boundaries.

In many cases, a fruitful methodology consists in integrating the discovery of lower-level causes with the investigation of higher-level features. For example, we might better understand a mental disorder such as depression by bringing together explanations from physiological, cognitive, and sociocultural levels. Such an approach might usefully explain why drug therapies may be successful in treating the disorder; why people with depression think differently about themselves and the world; and why depression occurs more frequently in certain populations. H. Atmanspacher and F. Kronz put this together in a nutshell, "an ontic description at one level serves as the basis for an epistemic description at a higher level, where it can be 'ontologized' and then provides the basis for proceeding to another epistemic description at yet another level."

In the rest of this book, I will show how the various disciplines in science may benefit from a multi-level approach, or more specifically a two-directional approach—a "bottom-up" method in search of upward causation, as done in a reductionist approach, as well as a "top-down" method in search of downward causation, as advocated by a holistic approach. A combination of these two methods may be more fruitful than limiting oneself exclusively to one specific method—although the latter approach may often be more pragmatic.

IV. Some Test Cases

In the following chapters, we will discuss some more specific examples of scientific explanations in various scientific fields that can be done in terms of downward or upward causation. I let you, reader, determine yourself whether you choose one of the two approaches, to the exclusion of the other, or decide to allow for both approaches in a supplementary way.

7. Physical Sciences

Although talk about reductionism and holism is not as common in the physical sciences as it is in the other sciences, there are both reductionist and holistic issues in the physical sciences as well. The physical sciences do not only include physics and chemistry, but also astronomy, meteorology, geology, and other earth sciences. In this chapter, we will only focus on physics and chemistry.

Let us start with chemistry. The case can be made that chemistry can be reduced to physics, that chemical laws can be derived from physical laws, and that molecules are aggregates of atoms. But although chemistry can be reduced to physics, there is also *emergence* in chemistry when molecules show new properties that fail to show up at the level of their atoms. The properties of oxygen (O_2) and hydrogen (H_2), for instance, do not account for the properties of water (H_2O); water owes its new properties to the relationships and interactions between two atoms of hydrogen and one atom of oxygen in a water molecule. Water consists of the same type of atoms as hydrogen peroxide (H_2O_2), but that does not mean that we should drink hydrogen peroxide instead of water when we are thirsty. The reason is that the way these molecules are put together matters. Interactions appear to be as "real" as their constituents, even in chemistry.

Another holistic case in chemistry might be the following example used by Timothy O'Connor. The joint action of multiple causes is not the sum of effects of the causes had they been acting individually in a case like this: NaOH + HCl → NaCl + H_2O (sodium hydroxide and hydrochloric acid produce sodium chloride and water). The product of this neutralization reaction, water and a salt, is in no sense the sum of the effects of the individual reactants, an acid and a base. Indeed, it is not very common among chemists to talk in terms of emergence and holism, but that does not mean this could not be an appropriate way of doing so.

Another interesting case can be found in what are called allosteric enzymes. Most allosteric enzymes have multiple coupled subunits; the activity of a subunit in the complex takes on a value that depends on the states the other subunits are in. An allosteric enzyme contains an active site, the location on the enzyme where it catalyzes its specific reaction, plus a second type of site, called an allosteric site. The allosteric site, through its binding of an effector molecule, influences the activity of the enzyme at the active site. Effector molecules may cause the enzyme to become more active or less active by redistributing the ensemble between higher affinity and lower affinity states. This allows most allosteric enzymes to greatly vary their catalytic output in response to small changes in effector concentration, thus making for an important tool in the regulation of cell metabolism. What is essential for our discussion here is the fact that these effects are properties of the whole system—the entire enzyme, that is—and not of single components, the subunits.

There are more cases in chemistry where one can find unexpected novelty emerging, as the so-called chemical clock of the Belgian physical chemist Ilya Prigogine illustrates. This case involves a linked series of reactions in which "red" and "blue" chemicals are synthesized or broken down through interactions involving other chemicals present in the solution. At fixed concentrations of the latter, an equilibrium might set in with a "purple" mixture of definite proportions. However, for some sets of circumstances, a remarkable rhythmic oscillation occurs, with the system alternating at regular intervals between red and blue like the ticking of a clock. What seems unusual coincidence at the microscopic molecular level can manifest a larger-scale organization at the macroscopic level due to the non-linear, reflexive character of the equations behind it. More in general, one could state that in case of chemical equilibrium, a single molecule does not "know" that it is in chemical equilibrium and must display a certain reaction-kinetic pattern in order to keep the equilibrium stable. In other words, the whole determines the average reaction-kinetic behavior of its molecular components.

The issue of emergence becomes even more pressing when we consider the first appearance of living systems on the early Earth. According to the scenario of Nobel Prize Laureate

Christian De Duve, life was preceded by a thio-ester world. This step was accomplished by thio-esters arising from the condensation of carboxylic acids with thiols. Thio-esters are capable of linking amino acids together into proteins, thanks to energy derived from the consumed thio-ester bonds. The small subset of polymers that "survived" in this process—through something like "molecular selection"—included a number of crude catalysts fore-shadowing present-day enzymes. Why are these "enzymes" so important? Well, a modern cell can only produce proteins with the help of DNA, RNA, and enzymatic proteins. The making of proteins is currently based on DNA and RNA, but DNA and RNA can only achieve this by using enzymatic proteins (see chapter 8).

That is where we run into a chicken-or-egg problem: no DNA and RNA without proteins, and no proteins without DNA and RNA. But the thio-ester world would solve this problem, because thio-esters are capable of linking amino acids together into proteins. In addition, scientists have discovered some special RNA molecules (ribozymes) which can also function as enzymes. Then there were molecules that possessed a site capable of specifically binding a given kind of amino acids; these were particularly favored by "molecular selection" again. Reciprocally, the amino acids that were recognized by proto-tRNAs were themselves selected to be used in a metabolism that depended on RNA binding.

This process can best be understood as an emergent phenomenon, because the simple precursors of living cells were composed of a self-assembled collection of molecules that by themselves were non-living, yet together led to new properties of self-maintenance, self-reproduction, and evolution. One might conclude from all the previous examples that a completely, exhaustive reduction of chemistry to physics is probably hard to maintain because chemical phenomena do depend on their physical components and physical laws, yet they are not uniquely determined by them

----

Let us next turn to physics itself. Although physics seems, at first sight, to be a perfect candidate for a one-level ontology—like in old, classic atomism—one could make the case that even here one-level explanations may be inappropriate, or at least incomplete. First of all, the presumed one-level identity of physics has already become a multiple-level issue for many physicists as they differentiate the molecular level, the atomic level, the sub-atomic level, and even further down. The question as to what counts as a part, and what parts are basic, can hardly be answered in the abstract, but is usually settled in a particular context of enquiry. Richard Healy mentions how atoms, for instance, would count as basic parts of hydrogen if it is burnt to form water, but not if it is converted into helium by a thermonuclear reaction.

So the question is: Can all physical systems and their properties be fully explained by and/or reduced to the properties of their subsystems? It is doubtful. As discussed earlier, a single water molecule does not have a temperature since temperature is defined only for groups of molecules interacting with each other. And depending on the temperature, these water molecules will form water vapor, liquid water, or solid ice. As Monica Anderson rightly remarks, "These three have very different properties. Could these different behaviors be predicted from the properties of individual molecules, such as van der Waals forces? It is difficult, and the reason we would even go looking for the connection is because we would have observed these emergent effects at the system level."

Another example is provided by the inability of detailed knowledge about the molecular structure of water to predict surface tension, which is another macroscopic phenomenon reflecting emergent behavior among water molecules. A bunch of water molecules in a glass will have surface tension, even though none of the water molecules themselves have surface tension in isolation. Something similar holds for the property of being "solid." This is not a property of individual atoms, but only emerges from interactions among many microscopic degrees of freedom.

Then Sean Carroll mentions the example of downward causation as found in snowflakes. Although they are made of water molecules acting according to atomic/molecular physics, the shape that they end up taking is highly constrained by the macroscopic crystalline structure of the snowflake itself. This structure would not have been visible if one were just thinking about molecules; apparently, the macroscopic structure has influenced the dynamics of the microscopic constituents.

Francis Heylighen adds the following example. It is well-known that snow crystals have a strict six-fold symmetry, but at the same time, that each crystal has a unique symmetric shape. The symmetry of the crystal, as a whole, is clearly determined by the physico-chemical properties of the water molecules which constitute it. On the other hand, the shape of the complete crystal is not determined by the molecules. Once a shape has been formed, the molecules in the crystal are constrained—they can only be present at particular places allowed in the symmetric crystalline shape. The whole of a crystal constrains or "causes" the positions of the molecular parts. This holistic terminology may look odd, nevertheless it seems to be entirely legitimate.

In many philosophical discussions, it is argued that the thermodynamic temperature of a gas is the mean kinetic energy of the molecules which constitute the gas. According to Thomas Nagel, for instance, this leads to a straightforward reduction of thermodynamic temperature to statistical mechanics. Such a rough picture, however, ignores some important details according to Carsten Allefeld. First, thermodynamic descriptions presume thermodynamic equilibrium as a crucial assumption which is not available at the level of statistical mechanics. Second, the very concept of temperature is basically foreign to statistical mechanics and is usually introduced in a phenomenological way. Another source of complexity is the large number of rather diverse systems to which thermodynamics can be applied, including not just gases and electromagnetic radiation but also magnets, chemical reactions, star clusters, and black holes. So these considerations might put into question to which extent thermodynamic properties are exclusively determined by the physical properties of the basic parts of thermodynamic systems.

Physics even allows for processes that exhibit some kind of "self-maintenance." Richard Campbell, for example, discusses a candle flame as a familiar example of a process that makes several active contributions to its own persistence. It maintains its temperature above the combustion threshold; it vaporizes wax into a continuing supply of fuel; and under standard atmospheric and gravitational conditions, it induces convection currents, thus pulling in the oxygen it needs and removing the carbon dioxide produced by its own combustion. Processes like these exhibit some form of self-maintenance.

The ability to be self-maintaining can be seen as an emergent causal power of the organization of the candle flame; for it cannot be explained simply as the physical resultant of the causal properties of its distinct constituents. Of course, in one sense its persistence is dependent upon its constituents—when the candle flame has burnt all its fuel, or it is deprived of oxygen, it ceases to be—but as long as the boundary conditions are fulfilled, it succeeds in maintaining its own process of burning. There is just an inflow of waxen and oxygen molecules, which are then consumed, and an outflow of carbon dioxide and water molecules (and other trace by-products), together with a release of energy. On the other hand, we cannot say what a candle flame is without mentioning its relations with external elements in its ambient situation. The very existence of the flame is a function of these external relations.

Let us consider next the example of superconductivity, where wholes determine parts rather than, or in addition to, parts determining wholes. It is a macroscopic event based on microscopic elements. The Nobel laureate physicist R. B. Laughlin and others have focused attention on a range of properties, insensitive to microscopic events, of various kinds of macroscopic matter, such as the crystalline state. They argue that the behavior of these properties is well-understood through high-level principles—while being inexplicable in lower level physical terms. Most of the physical properties of superconductors vary from material to material, such as the heat capacity and the critical temperature, critical field, and critical current density at which superconductivity is destroyed. On the other hand, there is a class of properties that do not depend on the underlying material. For instance, all superconductors have exactly zero resistivity to low applied currents when there is no magnetic field present or when the applied field does not exceed a critical value. The existence of these "universal" properties implies that superconductivity is a thermodynamic phase, and thus possesses certain distinguishing and emergent properties which are largely independent of microscopic details.

In this context, special attention should be paid to the Periodic System, because the organization imposed by Mendeleev's periodic table facilitated the discovery of concepts related to chemical bonding, electron shells, electronic orbitals, and ultimately quantum theory. To begin with, there is more to an atom of hydrogen than an electron and a proton; there is also the pattern of the relationship between them, and it is this pattern of the process, its organization, which is crucial to the emergent properties of hydrogen. Piotr Sadowski has showed us how Mendeleev looked at the elements not as separate, isolated components of nature, but as parts of a larger *whole*, possibly connected by shared properties. Elements placed in accordance of the value of their atomic weights present a clear *periodicity* of properties, which might be seen as a holistic concept.

As a matter of fact, the electron configuration or organization of electrons orbiting neutral atoms shows a recurring pattern or periodicity. The electrons occupy a series of electron shells (numbered shell 1, shell 2, and so on). Each shell consists of one or more subshells (named *s, p, d, f,* and *g*). As the atomic number increases, electrons progressively fill these shells and subshells. The orbitals of lower energy are filled in first with electrons, and only then the orbitals of high energy are filled. The electron configuration for neon, for example, is *1s2 2s2 2p6*. With an atomic number of ten, neon has two electrons in the first shell, and eight electrons in the second shell (with two in the *s* subshell and six in the *p* subshell). In periodic table terms, the first time an electron occupies a new shell corresponds to the start of each new period, these positions being occupied by hydrogen and the alkali metals. Since the properties of an element are mostly determined by its electron configuration, the properties of the elements likewise show recurring patterns or periodic behavior.

We could actually extend this discussion to the level of particle physics. As Péter Érdi notes, "There are certain interpretations of quantum mechanics that consider the perception of a deterministic reality, in which all objects have a definite position, momentum, and so forth, as an emergent phenomenon." It was Niels Bohr's view that one can meaningfully ascribe properties such as position or momentum to a quantum system only in the context of some well-defined experimental setting suitable for measuring the corresponding property. We cannot say that a photon is either a wave or a particle until it is measured, and how we measure it determines what we will see. The quantum entity acquires a certain new property—position, momentum, polarization—only in relation to its measuring apparatus, which could be interpreted in holistic terms. It is the emergence of a quantum entity's previously indeterminate properties in the context of a given experimental situation. The property did not exist prior to this relationship—it was indeterminate. Apparently, it depends on the interpretation we give to quantum mechanics whether we have emergence here.

Let us go back to the issue of ontological reductionism, which claims that all entities, from simple to complex, are ultimately based on simple, basic elements, whatever these may be (see chapter 6). It tells us that things turn out to be simpler than they appear. But what "simpler" means in science is not at all a simple question. Many ontological materialists would equate "simpler" to "lower, cruder, and more trivial"—brutish bits of matter, so to speak, mere "stuff."

In contrast, the theoretical particle physicist Stephen M. Barr would claim that things are not more coarse or crude or unformed as one goes "down" to the foundations of the physical world but rather more subtle, sophisticated, and intricate the deeper one goes. To explain his point, he uses the analogy of a large number of identical marbles rolling around randomly in a shoe box. If the box is tilted, all the marbles will roll down into a corner and arrange themselves into what is called the "hexagonal closest packing" pattern. This orderly structure emerges as the result of "blind" physical forces and mathematical laws. There is no "hand" arranging this structure. Physics requires the marbles to lower their gravitational potential energy as much as possible by squeezing down into the corner, which leads to the geometry of hexagonal packing. These two crucial features of the marbles—having the same shape and having a spherical shape—should be understood as principles of order that are already present in the supposedly chaotic situation before the box was tilted. So Barr concludes that the more we reduce to "deeper" explanations, the "higher" we actually go. This is because the preexisting order inherent in the marbles is greater than the order that emerges after the marbles arrange themselves.

Barr explains this further with the concept of symmetry, important in fundamental physics. When marbles arrange themselves into the hexagonal pattern, just six of the infinite number of symmetries in the shape of the marbles are expressed or manifested in their final arrangement. The rest of the symmetries are said, in the jargon of physics, to be spontaneously broken. So, in the simple example of marbles in a tilted box, we can see that symmetry is not popping out of nowhere. It is being distilled out of a greater symmetry already present within the spherical shape of the marbles.

Symmetry is one of the factors that contribute to so-called profound simplicity. "Grand unified theories," which combine the strong, weak, and electromagnetic forces into a single mathematical structure, posit symmetries that involve rotations in abstract spaces of five or more complex dimensions. Apparently, the simplicity to which ontological reductionism leads us is of a very paradoxical kind. It is a simplicity that is by no means simple-minded, but "profound simplicity," in Barr's own words.

Stranger and even profoundly simpler are super-symmetries. There is much reason to argue that super-symmetries are built into the laws of physics, and finding evidence of that is one of the main goals of the Large Hadron Collider outside Geneva, Switzerland. So we might say that the cosmos was at one point a swirling mass of gas and dust out of which has come the extraordinary complexity of life as we experience it. Yet, at every moment in this process of development, a greater and more impressive order operates within an order that did not gradually develop but was there from the very beginning.

One could even make the case that the reductionism-holism debate becomes futile if the world of subatomic particles is so intricate and complex that it also includes, at the lowest level, all the potentialities that unfold at so-called "higher levels" according to a holistic approach—including all the entities, properties, and relationships that we find "emerging" at higher levels. This would mean that holism is actually a phenomenon that is already enclosed in some kind of preexisting state at the lowest level of reductionism. The interesting part of this view is that it transcends the reductionism-holism debate, as holism would be "absorbed" by a very fundamental, but profound, form of reductionism.

For further reading:

Barr, Stephen M. "Fearful Symmetries." *First Things*, Oct. (2010).

Bohm, D. *Wholeness and the Implicate Order.* London: Routledge & Kegan Paul, 1980.

Leggett, A. J. *The Problems of Physics.* New York: Oxford University Press, 1987.

Placek, T. "Quantum State Holism: a Case for Holistic Causation", *Studies in the History and Philosophy of Modern Physics*, 2004, 35: 671-92.

Weinberg, Steven. Dreams of a Final Theory: The Scientist's Search for the Ultimate Laws of Nature. Pantheon Books, 1992.

8. Molecular Biology

Molecular biology has become one of the most thriving branches of the life sciences. As many of the molecular biologists in the 1950s came from physics, it is not surprising that they extended its classical approach to the study of living organisms. As Francis Crick put it, "The ultimate aim of the modern movement in biology is to explain all biology in terms of physics and chemistry."

Indeed, molecular biology has been very successful in dissecting organisms, organs, and cells into their molecular components. It has been so successful that some believe there is only one kind of biology—molecular biology. This is reductionism optima forma—undoubtedly backed by a great track record. One of its success stories is the discovery of DNA. It seems fair to say that DNA segments get passed on in much the same way that Mendelian genes are passed on.

However, the successes of the reductionist approach may have obscured several of its shortcomings. Despite the purported "hegemony of molecular biology," biological sub-disciplines that are focused on higher levels of organization have not disappeared and in fact remain prosperous. Besides, reductionism has not been implemented to the fullest. Ironically, no molecular biologist has ever found it particularly helpful to work with elementary particles at a level much farther down.

Nevertheless, the superiority, or even monopoly of the molecular level is still widely held among biologists nowadays. Often we can hear phrases like "The laws of genetics are the result of the structure of DNA," or "Living organisms exist for the benefit of DNA rather than the other way around." In these cases, it may be helpful to turn the tables by looking down from a higher level. This made the Harvard population geneticist Richard Lewontin comment as follows: "The laws of genetics are not the result of the structure of DNA, but rather DNA has been chosen by natural selection from among many molecules precisely because it fits the requirements of an evolved genetic system. DNA is only one out of several tactics for an evolutionary strategy." This is so because the coding of amino acids by specific triplets of bases in the DNA is not determined by any physical law. A given triplet might as well be translated into a multitude of other amino acids than the one operational in the organisms we know. But evolution happens to have selected one specific regime where the coding relation is unambiguously fixed.

According to the reductionist approach, on the other hand, the role of molecules should be at center stage in the life sciences. It has actually been given an "explain-it-all" role. But how far does this approach really take us? For example, a liver and a muscle are made up of the same material constituents—elements such as hydrogen, carbon, oxygen, and so on—acting on each other by the same basic forces. However, it is precisely their structures that differ and enable them to play different roles inside the organism.

In other words, their different roles do depend on these components, yet they are not uniquely determined by them. Other factors have to be taken into consideration. That's where a holistic approach could help us out—without denying, though, what a reductionist approach has revealed to us. To paraphrase the late biologist Barry Commoner, there are two different ways of looking at something like "life": either in the reductionist way of "DNA is the secret of life," or in the holistic way of "Life is the secret of DNA." One does not preclude the other.

No matter which side we take, one thing remains rather undisputed: Although higher level entities may constrain processes at a lower level, the laws of nature at the lower level cannot be overturned. As chemistry cannot defy the laws of physics, biology has to be consistent with the laws of both physics and chemistry. On the other hand, biological systems do have a history that physical systems do not—they store *information* and process information at a very fine, atomic/molecular level. Norbert Wiener made clear to us that information is neither matter nor energy, but it needs matter for its embodiment and energy for its communication.

Right from the time of death, the corpse begins to disintegrate. It literally disappears, unless the natural process of disintegration is artificially suspended—by freezing, embalming, or placing it in formalin. Obviously, there is something to being a living organism that is more than the assemblage of atoms and molecules. Considerations like these would call for a radically different ontology. In the words of Richard J. Campbell, "Biological systems—including humans—are not substantial entities ('things' in the thick sense) whose constituents are cells (smaller things), which in turn (after a few more reductions) are constituted out of elementary particles. Instead, they are open, organized action systems, in essential interactions with their environments, such that we cannot say what they are without taking those interactive processes into account."

These systems can maintain their stability only as a result of their interactions with their surroundings. Their very existence as well as their persistence are dependent upon their relations with external factors in that environment from which they keep drawing sustenance. In this respect, biological systems are somehow comparable to candle flames (see chapter 7). As complex organizations of processes, they persist only so long as they are able to maintain appropriate interactions with their environment, by which to sustain their existence. They eat and drink, perspire and excrete, and in many cases, breathe. When they stop those activities, they die.

From a holistic perspective, we should therefore take a different look at all the molecules that molecular biology has discovered and analyzed based on a reductionist approach. They have been treated as the magic molecules determining life from "the bottom up." However, all these molecules (DNA, proteins, fatty acids, etc.) are fabricated through internal processes within the cell itself—at a higher level, that is. They are generated through a web of interactions of the whole system. That downward causation occurs is considered a fact by most biologists; but how to understand the phenomenon is the contentious issue. One of the first questions is this: Is the structure of proteins (structural proteins and enzymes) and nucleic acids (DNA and RNA) completely determined by its components? Let us briefly discuss this issue for proteins first.

Functional proteins are three-dimensional, folded structures composed of amino acid components linked together into a linear chain. Reductionism would study how the folded protein is a result of the amino acids it is composed of. However, such a reductionist approach would be rather limited, since new studies have emphasized, among other factors, the necessary role of additional elements, including other folded proteins to assist in the proper folding of newly generated linear amino acid chains. The production of such molecules is staggeringly complex and speedy. When 200 million of the 25 trillion red blood cells die each second, the 100 million hemoglobin molecules in each cell must be replaced immediately. With the order of a few thousand bytes of information in each hemoglobin, this is 10 thousand x 100 million x 200 million = 2 x 10^{20} bits of information per second—a million times more information processing than today's fastest computer CPU.

A similar point can be made regarding DNA molecules. If the sequence of DNA were fully predetermined by chemical bonds, DNA would not be able to store or convey information. Purely physical processes would be bound to lead to a destabilization of information. A mixture of chemical substances will follow the second law of thermodynamics and ultimately revert to the most probable state of chemical equilibrium. Yet, DNA bears information! But like with a sheet of paper or a computer memory chip, its chemistry is irrelevant to its content.

DNA is a "neutral" carrier of information—in a sense independent of its chemistry and physics. Its information is transferred through physical and chemical carriers—but it is not *specified* by physical and chemical forces. Seen from a purely physico-chemical perspective, any order in the DNA sequence is possible. Therefore physics and chemistry cannot specify which order will in fact succeed in functioning as a code. As George Gilder puts it, "Information is defined by its independence from physical determination: If it is determined, it is predictable and thus by definition not information."

This makes many biologists realize that the analytic, downward approach of studying DNA does not get us very far, or at least not far enough. But once we go for the upward, synthetic approach, new horizons open up. It turns out that DNA is part of a much wider context. So, instead of focusing downward, we might need also an upward approach. The reference to a *code* is not on the same explanatory level as the laws of physics and chemistry. In the words of Marjorie Grene, "The explanation through reference to a code is a hierarchical explanation that adds additional constraints on the operation of the system." By disorganizing the code, the organism may die, although there is nothing "wrong" in a physical or chemical sense. Just as it would be absurd to claim that an understanding of Shakespeare's works can only be achieved by studying the English alphabet, so it would be silly to say that the DNA code can only be understood by studying nucleotides.

Take the following example. In the thirties, it was discovered that fermentation could be maintained in a cell-free extract—certainly a success story for the reductionist approach—but since this process had destroyed cellular material, it could not be determined where in the cell the various reactions take place. Studying the relevant *organization* inside the cell—inside the whole, so to speak—required new instruments such as the electron microscope and the ultracentrifuge. Once the organization, including compartmentalization, of a cell had been acknowledged, it became clear that DNA, for instance, is "useless" in itself, unless it is part of a larger system which includes enzymes and other cellular components located in the "proper" cellular compartments.

It is at this higher level that cell biologists feel at home; they have discovered that DNA is only a small link in a complex process of protein synthesis (including ribosomes, RNA, and enzymes). And this process, in turn, has been integrated into an even more complex system at the level of the organism. Without being part of a larger system, DNA just does not do what it does in an organism. This calls for macro-to-micro explanations. In the words of George Gilder, "The biological cell is no longer a 'simple lump of protoplasm' as long believed but a microcosmic processor of information and synthesizer of proteins at supercomputer speeds."

Those who still focus exclusively on the downward approach, which reduces everything in biology to the "power of DNA," should at least be aware that DNA can never do anything on its own. First of all, it is not even capable, as some still believe, of self-replication—that is, making copies of itself on its own—for DNA is manufactured out of small molecular bits and pieces by the use of an elaborate cell machinery that is made up of proteins. If DNA is put in the presence of all the pieces that will be assembled into new DNA, but without any protein machinery, nothing happens. It is actually the presence of many other components that makes sure old DNA strands are replicated into new strands. For this reason hemoglobin, for instance, cannot be reduced to a certain DNA sequence, for it is only in combination with other factors that this sequence can produce hemoglobin.

Second, although the DNA code may have coding units—so-called codons, composed of three nucleotides—the "meaning" of these codons is not entirely fixed but depends also on the surrounding code—at a higher level, that is. This is comparable to regular language where certain words receive a meaning from their context. The Harvard geneticist Richard Lewontin likes to mention the DNA sequence of the nucleotides G-T-A-A-G-T as a case in point. Usually the cellular machinery will read this as a two-codon instruction to insert the amino acids valine and serine into a growing protein chain. However, sometimes this very same sequence is read as a code which regulates the expression of a neighboring gene, and at other times it acts as a blank separating two different DNA sequences. In order to find out which interactions cause which interpretation of the very same DNA sequence, we need a more holistic approach with top-down causation and macro-to-micro explanation.

Third, DNA on its own does not even produce proteins. The role of DNA is to provide a specification as to how amino acids are to be strung together into proteins by some synthetic machinery, but this string of amino acids is not yet a protein. To become a protein with physiological and structural functions, certain parts may have to be excised and then the string must be folded into a three-dimensional configuration that is only partly based on its amino acid sequence, but depends also on the cellular environment and other special processing proteins.

Commercial insulin for diabetics makes a case in point. Not too long ago, the DNA coding sequence for human insulin was inserted into bacteria, which were then grown in large fermenters until a protein with the amino acid sequence of human insulin could be extracted. But it turned out amino acid sequence alone does not determine the shape of a protein. The first proteins harvested through this process did have the correct sequence, but were physiologically inactive. Apparently, the bacterial cell had folded the protein incorrectly. Somehow the DNA for a protein does not "know" how to fold a protein, so as to make it work. This may happen more often than initially thought. Amyloids, for instance, are insoluble protein aggregates that arise from inappropriately folded polypeptides naturally present in the body. These wrongly folded structures may even play a role in various neurodegenerative disorders, including Alzheimer disease. In other words, DNA surely does not "know" everything on its own.

All of the above pleads for a downward approach in addition to a reductionist approach. One could argue, though, that seen from a cellular level, all the previous mechanisms related to DNA at a lower level are the result of reduction, making for various forms of upward causation. But this does not obliterate the fact that the information which DNA carriers is not determined by its constituents—through upward causation, that is—but by a downward causation found at higher organizational levels.

As said before, any arrangement of the four bases in DNA is compatible with the laws of physics and chemistry—otherwise it could not exist at all. But in terms of biological information, it is just the particular order that makes all the difference. Nevertheless, there is more to it. Not only must the scientist recognize the existence of a code, but in some sense the organism must "recognize" it too. This requires a *two*-level process, for reading the code can succeed or fail—which in turn leads to a third level, the formation of an organism if the code succeeds. So even molecular biology—not to mention other branches of biology—can hardly be seen as a one-level enterprise.

The basis of phenomena like these is that a "coordinating" organization is somehow channeling and harnessing the contribution of the parts. This is what Michael Polanyi called a biological constraint (see section III). Most molecular biologists take this "higher" organization as a given they can ignore, but that does not mean it doesn't exist and play a crucial role. At higher levels—which are more intricately structured—more boundary conditions seem to reign. Whereas physical constraints can be of any kind—e.g. any speed or position or temperature—biological constraints represent a specific selection of constraints from a multitude of physically equivalent alternatives. What then is this selection accomplished by? The simplest answer for now is: by natural selection, which determines at a higher level which physically equivalent alternatives are better or worse for survival (see chapter 9).

Put differently, the lower level leaves open boundary conditions to be specified by the laws of the higher level. DNA is not simply some chemical molecule but rather a biological structure that can be understood and explained in terms of programs and blueprints. In other words, to explain the working of DNA, we do not just look to chemistry and physics, but also to engineering and information technology. The *message* of a code may be analyzed into its particular constituents, but it cannot be reduced to them. Talking of "code" is actually "design" talk, which is foreign to physics. (More on design talk in chapter 11.)

One word of caution, though, about this "information" terminology. Peter Godfrey-Smith points out the following. Whereas the information held by text can be instantiated in many different physical forms—written in ink or pencil or chalk, on paper or on slate, in electronic characters on a screen, or not written at all but spoken—the causal properties of DNA, including the effects it has on an organism's development, require it to be in the physical form that it is and no other. A text printout of C's, G's, T's and A's would not do the job. So the analogy of concepts such as "information" and "code" borrowed from the informational sciences may be helpful but also deceiving at the same time.

For further reading:

Brigandt, Ingo and Alan Love. "Reductionism in Biology." *The Stanford Encyclopedia of Philosophy*, Edward N. Zalta (ed.), 2012 http://plato.stanford.edu/archives/sum2012/entries/reduction-biology/.

Commoner, B. "Roles of Deoxyribonucleic Acid in Inheritance." *Nature*, 202 (1964): 960-968.

Godfrey-Smith, Peter. *Philosophy of Biology*, Princeton: Princeton University Press, 2014.

Polanyi, Michael. "Life's Irreducible Structure." *Science* 160 (June 1968): 1308–1312.

9. Population Genetics

As we discussed earlier (see chapter 1), good scientists are those able to demarcate their area of investigation by limiting themselves to factors that are relevant to what they are studying, and by keeping strict control over factors that might interfere with their search. To put it more technically, scientists create a simplified model of what they are studying. They reduce complex entities to a manageable model related to an analyzable problem as a successful way of doing research.

One of the successful moves the first geneticist, Gregor Mendel, made was that he did not attempt to study the offspring of all organisms and all their characteristics at once. Instead, he selected only a few individual traits of specific organisms, pea plants, and studied them in more detail. Mendel knew how to select and isolate things by limiting himself to a small selection of specific details. Besides, he did not study a single organism, but a large number of them to allow for randomization and statistical analysis. His successors have continued this approach ever since.

Population geneticists do too. Take the case of a gene-pool model—a model very popular in population genetics. It is based on the following methodological reduction: A population is reduced to a reproductive community of organisms, an organism is reduced to the outcome of a genotype, and the genotype is reduced to a summation of certain genes (which carry variants called alleles). Population geneticists can use this model to study the effects of inheritance when exposed to natural selection.

The simplest models of a gene pool tend to focus on a single specific gene with its various alleles; they simulate how selection pressure on this particular gene may change allele frequencies by selectively promoting reproduction of certain alleles of this particular gene. To put it briefly, the gene pool model simulates the "survival battle" of alleles for one particular gene. So we could end up with the statement that a population "is nothing but" a gene pool, in which the allele frequencies change due to natural selection.

Let us analyze this concept in a more technical, mathematical way. Mendel's first "law" describes how genes are passed from one generation to another. If the gene pool is composed of two alleles, say A and a, and the frequency of allele A is p, then the frequency of allele a must be $1-p = q$. From this, the frequency of the different genotypes can be deduced as follows: The frequency of genotype AA is $p \times p = p^2$, the frequency of genotype aa is q^2, and the frequency of Aa plus aA is $pq + pq = 2pq$.

Further mathematical analysis shows that if all genotypes mate at random with one another, then their frequencies in the next generation remain stable. Conclusion: If the frequency of AA was p^2 in the previous generation, it will remain p^2 in the generation(s) to come. Similar calculations show that the same holds for the other genotypes as well. It is the gene-pool model that makes this mathematical approach possible and fruitful. This has led to the so-called Hardy-Weinberg law (or theorem), which states that, all other things being equal, allelic frequencies and genotypic ratios remain constant from generation to generation.

However, in real life, "all other things" are never equal. First of all, mating and reproduction are not really random. Second, there is mutation from one allele to another. Third, there is immigration and emigration. Fourth, alleles may undergo a "random walk" process, called genetic drift. Fifth, and most importantly, certain alleles have a better chance to reach the next generation than others—which means that they differ in "fitness." One of the two alleles may have a selective disadvantage, which is called a negative selection pressure, whereas the other one undergoes a positive selection pressure. As a result, the frequencies of alleles may very well change in future generations, instead of staying the same if reproduction were completely random. It is the gene pool model that helps biologists simulate and analyze this process.

One word of caution. Although the gene-pool approach seems rather mathematical and mechanical—similar to Newtonian theory in physics—it is also *teleological*, for it has an endpoint, survival, which is to be understood as the goal of what leads up to it. Survival is a concept that does not exist in mathematics and physics, making biology a science different from physics; biology refers to phenomena that are design-like, whereas physics does not. It is thanks to natural selection that organisms have a history, whereas physical systems typically do not. It is also thanks to natural selection that organisms have features with successful effects, which is foreign to physics. Physics in general rejects "for the sake of" statements—for instance, when the temperature rises, gases do not expand "in order to" keep the pressure constant. Yet, life scientists like to say, and are entitled to say, that the heart pumps "for the sake of" circulating blood.

While gene-pool selection seems a purely mathematical or statistical issue, Darwinian selection is environment-based and causal. But these two approaches can be fused under one and the same concept, *fitness*. When done so, we should distinguish between F-fitness and D-fitness. If we fail to do so, the idea of natural selection could easily be attacked by arguing that it amounts to a tautology—that is, a statement which is true in every possible interpretation. Critics of that idea use an argument that goes like this: "Who survive? The fittest! Who are the fittest? Those who survive!" However, we don't have to take this attack seriously, for it is based on some terminological confusion. Biological fitness has actually a double meaning: It refers to the role of an organism as the subject, or cause, of reproduction, but also to the role of an organism as the object, or effect, of reproduction.

This ambiguity also affects the fitness concept: Darwin's concept of fitness—let us call it D-fitness—is potential reproductive success (a cause), but there is also the concept of fitness in the sense of actual reproductive success (an effect)—let us coin this version F-fitness, which is the way the statistician Robert Fisher and some population geneticists use it. Apparently, the slogan "survival of the fittest" is not a tautology if we take fitness as D-fitness, or reproductive capacity. The principle of natural selection (or survival of the fittest) asserts that those organisms that are *potentially* successful in reproduction (D-fitness) are more likely to be also *actually* successful in reproduction (F-fitness). Thus having a good design does matter in evolution. The gene-pool model can demonstrate it.

It looks like all of this is highly reductionist in its approach. Thinking in terms of gene pools appears to be particulate and reductionist, for it works with genes and genotypes instead of organisms. However, we may easily be overlooking some holistic elements. One of them is that we have here some form of hierarchical thinking in so far as it is the laws of the collection, not of its elements, which are being investigated. In addition, there is a holistic approach here that uses downward causation in so far as a population of individual organisms is constrained by processes of natural selection occurring at a level above the level of the gene-pool model. Selection in the gene-pool model is reduced to the lower level of genes (genotypes), but adaptations occur at the higher level of organisms (phenotypes). Natural selection does not "select" genotypes but rather phenotypes. If we do not acknowledge this, we could run into the following problem.

Once we lose sight of the context of higher levels, it might seem like the actual unit of selection is located in the gene (through its alleles). When that happens, methodological reduction—downwards to the level of genes and their alleles—becomes an ontological issue so that the "real" units of operation are presumed to be genes and their alleles. What is wrong with that? Why can genes and alleles not be the "real" units of selection? The answer is that in the words of the late Harvard biologist S. J. Gould, "selection simply cannot see genes," because recessive alleles—which do not come to expression in the organism—cannot be subject to selection, since selection just cannot "detect" them. Therefore, alleles may very well be considered the units of selection in terms of the gene-pool model, but not so in the "real" world. In addition, selection may not even "see" an individual gene, but only a genome at a higher level—a totality of genes that have an impact on each other within the setting of an organism (see chapter 10).

Another problem of ignoring the existence of higher-level entities is that without this higher-level context, the gene-pool model cannot take into account organisms at a higher level either. This may have unexpected ontological implications: Organisms are no longer considered "real," only genes presumably are. If that were true, organisms can at best only be vehicles that carry genes to the next generations. This view may lead to all sorts of scientific gibberish such as claiming that alleles are "essentially" elements that are "selfishly" engaged in a constant "survival battle" with other alleles.

However, genes cannot be selfish or unselfish, any more than atoms can be jealous. Take it as playful metaphorical language, but nothing more. Talking about "selfish genes" makes only sense in the context of the gene-pool model, but detached from that model, it makes no sense at all.

Or consider this kind of language: All organisms are essentially mere "survival machines" for their genes. This is something similar to claiming that a chicken is essentially the egg's way of making another egg. Each time, the deceiving keyword here is hidden in terms like "essentially"—suggesting that the model tells the whole story and can operate only without a higher-level context. Treating a population as if it is a gene pool does not transform it into a gene pool—that would be methodological reductionism turned into ontological reductionism. Some may say there is nothing wrong with such a move, but at least it should be acknowledged that this is no longer a scientific step but a philosophical one.

This issue has become rather prominent since Richard Dawkins' argument for the primacy of genes. Entities basic to his theory are "replicators"—which are things with a high precision of replication. Cells, organisms, and populations are supposedly mere "vehicles" to be used by these replicators or genes. Somehow every organism is supposedly manipulated by a set of genes, each one seeking to have the vehicle make as many gene copies of itself as possible. This has Dawkins claim, "Living organisms exist for the benefit of DNA rather than the other way around." It is a new version of a much older remark that genes run the individual organism in their own "interest." Later on, however, Dawkins made it clear that it is only in a certain sense that genes are more primary than organisms. They are only primary as units of replication, but not necessarily as units of selection.

This was a much-needed clarification. Genes may very well be units of replication, as long as we keep in mind that they cannot even replicate and reproduce without the products of other genes. But genes are definitely not units of selection. The organism itself appears to be the most likely candidate as a unit of selection, for the simple reason that organisms are the only beings to procreate and die. The organism appears to be the unit of selection in the same way as the population is the unit of evolution. Calling genes the units of selection makes sense in the context of a gene-pool model, but detached from this model, it makes no longer sense.

Let us round off the gene-pool model discussion. Since this model can also be used for multiple genes, or even the entire set of genes in a population, some people call this kind of approach disrespectfully "bean-bag genetics"—a bag of genes in a bag of chromosomes in a bag of cells in a bag of organisms. It is reductionism optima forma—often helpful when based on methodological reduction, but deceiving when it becomes ontological reductionism. That's where science can easily slip into science-fiction.

For further reading:

Brandon, R. and R. Burian. *Genes, Organisms and Populations: Controversies over the Units of Selection.* Cambridge, MA: MIT Press, 1984.

Dawkins, Richard. *The Selfish Gene.* Oxford: Oxford University Press, 2006, Chap. 12.

Gould, Stephen J. "Caring Groups and Selfish Genes." *Natural History*, 86, 12 (1977), 20–24.

10.   Genome Genetics

Before we get started, we need to find out first what the difference is between genome, genotype, and phenotype—if there is any. The shortest possible answer would be that all three reside at different organizational levels. However, the basic concepts of genome, genotype, and phenotype are not defined in a satisfactory manner within the biological literature. Not only are there inconsistencies in usage between various authors, but even individual authors do not use these concepts in a consistent manner within their own writings. One study by M. Mahner and M. Kary found at least five different notions of genome, seven of genotype, and five of phenotype current in the literature. Perhaps we can create a bit more, or at least some basic, terminological clarity here.

First there is the level of the *genotype*. Usually the term genotype refers to the genetic makeup for a specific characteristic under consideration—characteristics such as albinism, colorblindness, sickle cell anemia, blood types, etcetera. A simple, though outdated, genetic notation for the genotype of blood type AB would be something like AB or I^A/I^B. (The new notation for I^A would be: transferase A or alpha 1-3-N-acetylgalactosaminyltransferase (A3GALNT); and for I^B: transferase B or alpha 1-3-galactosyltransferase (A3GALT1)).

Besides, the term genotype has also been used as a collective term for several, or even all, individual genotypes of an organism combined. A simplified and outdated genetic notation for the genotype of the combined blood types AB and Rhesus-positive could be something like AB+- or $I^A/I^B/D/d$. In this latter sense, a genotype carries genetic instructions for more than one characteristic of an organism. In either case, individual characteristics of an organism are connected to specific genes. Such genes are sometimes called structural or expressive genes—to distinguish them from DNA sections that do not create an end-product.

In addition to the term genotype, there is the term *genome*. Let us stress first that from a reductionist viewpoint, it does not make sense to distinguish these two terms, since they are both based on the same building blocks—nucleotides stringed together in DNA or RNA. But in a holistic context, it is important to make the distinction. The concept of genome has become very prevalent through the *Human Genome Project*, which started as an international scientific effort to map all the genes on the 23 pairs of human chromosomes and, to sequence the 3.2 billion DNA base pairs that make up the chromosomes. Begun in 1990, the project was largely completed in 2000 when 85% of the human genome was decoded, and ended in 2003 with 99% de-coded; detailed analyses of all the pairs were published by 2006.

This project also revealed that we share 98% of our human DNA sequence with chimpanzees, and 40% with lettuce, for instance. This caused quite some controversy in interpretation. Some claimed that we are "98% chimp"—which is reductionism writ large—while others said that we are no more "98% chimp" than we are 40% lettuce. The biological reality is we are not them, and they are not us. As the *Human Genome Project* states:

In February 2001, scientists working on the project published the first interpretations of the human genome sequence. Previously, many in the scientific community had believed that the number of human genes totaled about 100,000. But the new findings surprised everyone: both research groups said they could find only about 30,000 or so human genes. This meant that humans have remarkably few genes, not that many more than a fruit fly, which has 13,601 (scientists had decoded this sequence in March 2000). This discovery led scientists to conclude that human complexity does not come from a sheer quantity of genes. Instead, human complexity seems to arise as a result of the structure of the network of different genes, proteins, and groups of proteins and the dynamics of those parts connecting at different times and on different levels.

In short, a genome is an organism's complete set of DNA, including but not limited to all of its expressive genes. This raises the question if there is more to a genome than genes— or is a genome merely the sum total of all individual genes. The answer is that the genome also includes repetitive and non-coding sequences of DNA. This certainly adds to the size of a genome. Genome size is the total number of DNA base pairs in one copy of a haploid genome. An organism's complexity is not directly proportional to its genome size; some single cell organisms have much more DNA than humans.

Neither is variation in genome size proportional to the number of (functional or expressive) genes. Whereas the house mouse, *Mus musculus*, has 2,700,000,000 base pairs (2.7Gb), *Homo sapiens* has 3,200,000,000 base pairs (3.2Gb). The amount of non-coding DNA varies greatly among species, but in the human genome up to 98% is considered non-coding DNA.

We are discovering more and more about these "extra" parts in the genome. Their function is still very much up for debate. Initially, a large proportion of non-coding DNA had no known biological function and was therefore sometimes referred to as "junk DNA," particularly in the lay press. This is, however, a very misleading term, because it has become more and more evident that this DNA is not as useless as initially thought. This alleged "wasteland" has proved to be a repository for a variety of functions that are part of normal, and even critical, cellular processes. For that reason, it is much better and safer to speak of "non-coding," "neutral," or "silent" DNA, to distinguish it from expressive DNA (genes). Just because the function of a specific region is not known does not mean that it has no function at all. Absence of evidence is not evidence of absence. What we are doing here is like reading the lines of a computer program without really knowing yet the full syntax and semantics of the programing language.

Some geneticists consider the genome to reside at a level *below* that of genes and genotypes—just one long sequence of DNA—but such a view may prevent them from missing out on some important new insights. The level of expressive genes and genotypes can hardly explain all it is supposed to explain. Admittedly, there are genes for the various characteristics of eyes such as eye-color, vision pigments, color-blindness, etc., but it is hard to believe that a summation of all these features and their underlying genes combined makes for "having eyes." It seems more likely that seen from a higher level, the genome is more than a mere sum of genes and genotypes that determine the various characteristics of an eye. These latter genes do not generate eyes but only certain eye features. Something else is needed in addition. A large part of the genome may act like "switches" that can turn expressive genes on or off.

The *Human Genome Project* report seems to convey the same message too, when it stated that "human complexity seems to arise as a result of the *structure* of the network of different genes, proteins, and groups of proteins and the dynamics of those parts connecting at different times and on different *levels*" (italics are mine). Seen this way, the genome would reside on a level *above* the level of genes, including control over how to process the blueprint stored in expressive genes. To use "programming" terminology, a gene does what the program of the genome dictates; but it is not the other way around—the program does not do what a gene dictates. Seen this way, a genome is supposedly more than a mere aggregate of genes; it operates at a higher level with the help of "switches," etc. A genome contains also instructions which regulate what to use from its archive of expressive genes when, where, and how. Certain parts of the genome must influence the entire organism, not just single traits.

If we take the genome merely as a long-winding sequence of nucleotides, it would be reduced to a lower-level entity. This would be equivalent to treating Shakespeare's plays as a massive sequence of characters. But if we take the genome as a very structured entity with a structured sequence of nucleotides—like we would treat Shakespeare's work—it would rise to a higher level. Compare this with the use of language. We mentioned earlier how Monica Anderson uses the example of a word such as "like" that can be used as a noun, verb, adverb, adjective, preposition, particle, conjunction, and filler—depending on the context in which it is used. Something similar seems to hold for the DNA sequence G-T-A-A-G-T, which we discussed earlier (see chapter 8). There must arguably be something above the gene level that regulates how this sequence actually is executed. As put metaphorically by the systems biologist Hiroaki Kitano, "Identifying all the genes and proteins in an organism is like listing all the parts of an airplane. While such a list provides a catalog of the individual components, by itself it is not sufficient to understand the complexity underlying the engineered object. We need to know how these parts are assembled to form the structure of the airplane."

If this is true, then the genome is not a randomly generated sequence of nucleotides, but most likely a highly structured and organized entity. What is even more striking is the fact that the genome contains repair mechanisms to combat random disruptions. When random changes such as mutations do occur, they can disrupt this delicate structure of the genome. If the genome is indeed more than the sum of all its expressive genes, mutations may affect more than genes, namely the switches and constraints that form part of a complicated program. Changes in the regulatory part of the program may have much more drastic effects than changes in genotype. To put this again in computer terminology, branching statements such as "If..., Then..." work differently than the regular statements they include.

Applying this to genetics, we need to look for elements in the program that the genotypes are submitted to. There are, for instance, variations in the DNA of the genome outside the coding regions of expressive genes that do not affect the chemical structure of proteins but rather affect the rate of their synthesis. That may be part of the holistic role of a genome. The higher level of the genome may determine if, when, and how the lower-level of genotypes will be expressed.

Regulation in the genome is actually poorly understood as yet, but it is known that a combination of activation and suppression switching operations is involved. A typical gene is, on average, associated with some ten switches, combinations of activation and suppression by other proteins that lock onto the switch sites, so a complex network of interacting proteins is almost always required to start and stop a particular gene activation. The regulatory genes (or "transcription factors") will act in many different ways at different locations during the developmental process. The most powerful of these—called "tool box genes," including the homeotic or *Hox* genes—can activate very complex top level building blocks and then trigger, for instance, the complete development of an entire limb at a certain location. These triggers can perform very different construction jobs within the same organism or in different types of organisms.

As said before, genes come in two main forms, expressive and regulatory (however, some perform both functions). The expressive (or structural) genes are those that actually create the cell proteins for structure and metabolism; these we can regard as the "low level" genes. But before a gene can be expressed, it needs to be "switched-on," and this is a function of the genetic regulatory system, the "higher level" control process for the cell. From a reductionist point of view, such distinctions do not even make sense since we are dealing in all these cases with the same building blocks of nucleotides.

Another reason why the genome seems to belong to a higher level than genes is the discovery that genes are not very well defined physical units. Here is why not: (1) There are examples of "genes" that are split into pieces. The transcript from one piece is joined to the transcript from another one to produce a functional RNA. (2) Since some "genes" overlap, a single stretch of DNA can be part of two or even three genes. (3) In some cases the primary transcript is extensively edited before it becomes functional. It may even happen that nucleotides are inserted and deleted. In such cases, the original "gene" requires the assistance of other "genes" to ensure a functional end-product. (4) There are units of DNA containing a cluster of genes under the control of a single regulatory promoter. The genes are transcribed together into one mRNA strand, after which they are either translated together, or they undergo trans-splicing to create several strands of mRNA, each of which encodes its own single gene product.

Because of all the above reasons, it has become very hard to come up with a comprehensive definition of a "physical" gene. The boundaries of a gene have become rather blurred and make only sense within the larger setting of a genome. Most biologists would agree with all of this without explicitly recognizing that this is actually a form of holistic thinking.

In addition to genome and genotype, there is also the concept of *phenotype*. Again, it can be used for a particular characteristic, but also for a series of characteristics. It resides at the next level up—above the genotype level and above the genome level. The phenotype takes shape by integrating the "contributions" coming from the genome and genotype level as well as those from the environmental level. Developmental processes take off on the road from genotype to phenotype. They build on genome and genotype but are certainly not fully determined by them. That is the point where emergent properties may emerge again.

A case in point would be the development of the human brain. As G. M. Edelman notes, it has too much architectural complexity for it to be plausible that genes specify its wiring in detail; therefore developmental processes carry much of the burden of establishing neural connections. It is at this level that environmental input becomes a determining cofactor (see chapter 13). Learning processes and other higher-level factors contribute to shaping the brain structure. This phenomenon has become known as the "plasticity" of the brain.

Another example would be the development of gender differences. A difference in gender entails much more than a difference in biological characteristics—namely also differences in behavioral traits, social roles, and cultural expectations that come with being a man or a woman. Early on in human development, parents as well as society take on a molding role at a level above the level of genome and genotype. For example, as soon as parents know their child is a boy, they treat it as a boy, which makes the child consider himself as of the male gender. So this raises the question whether differences in gender are only the outcome of differences in sex, which in turn are supposedly based on differences in genes—or is there perhaps much more going on.

Apparently, the distance in a case like this between genotype and phenotype is a rather extended one. First of all, there are also sex hormones involved; however, genes do not produce sex hormones, but rather the enzymes that in turn produce hormones—which is at least a two-step process. Even if one would argue that genes affect hormones, and that hormones affect the brain, and that the brain affects our behavior, then it should be noted that this behavior, in turn, can affect the brain again. A similar phenomenon is well known from sports too: Strong muscles benefit those who play sports, but in turn, playing sports greatly benefits the development of the muscles.

In other words, it is not only the genes, the hormones, and the brains that shape our behavior, but everything that we see and hear around us, plus all the dreams, ideals, hopes, plans, and expectations we foster in our minds. All these factors, coming from a level above the level of genome and genotype, have an impact on the way humans develop themselves—their phenotype, that is. What seems to be "in-born" may in fact very well be "in-printed" or even "self-taught." Raising children is more than giving them a genome. That is where nature and nurture meet. However, very often, or maybe even always, it is nearly impossible for scientists to tell exactly what part is "nature" and what part is "nurture." And yet, most of the time, there is in fact a nurture part—for the simple reason that a phenotype is not only the product of a genome.

Usually it is assumed in the nature-nurture debate that it all starts with "nature," but that "nurture" may interfere during the rest of the developmental process towards a phenotype. True, genetics assumes that there is a one-directional flow from genotype to phenotype—in other words, the genotype can change the phenotype, but supposedly the phenotype cannot change the genotype. While it is possible for genes to change someone's behavior, there is no evidence to suggest that one's behavior can change the genes. Obviously, bodybuilders do not produce descendants who have bodybuilder features already built into them. When dentists pull wisdom teeth, they are not pulling genes, so people from the next generation will struggle again with their wisdom teeth. In other words, these "external" changes did not make it to the genes.

However, this assumed one-directional flow—upward causation without downward causation—is not cast in iron; it could very well be the result of a biased reductionism. There are some indications that environmental influences can etch chemical modifications in DNA. Subtle chemical markings on the DNA of stem cells recovered from the umbilical cord blood of babies, for instance, were found to be significantly different for different body sizes, as Francine Einstein and John Greally, and others, have reported. So there might be influences in the uterus that do affect genetic mechanisms. A rigid form of reductionism could miss such clues if there are any.

It is still not very well understood how environmental input can cause downward causation towards the level of genotype or genome. Perhaps this is the right moment to make an important distinction between the words "hereditary" and "genetic." Heredity is passing on traits to offspring, but this can be a matter of either genetics or upbringing. Some traits may be hereditary but not genetic; owning lots of money, for instance, may be hereditary but certainly not genetic. Some other traits, on the other hand, may be genetic but not hereditary, because they are the result of genetic changes limited to a specific cell line of the organism (unless the change occurred in the cell line of reproductive cells). A mosaic, or mosaicism, is genetic but not hereditary; it denotes the presence of two or more populations of cells with different genotypes in one individual. For instance, individuals with two different eye colors or with a patch of white hair probably had a mutation in certain cell lines, so they will not pass this on to the next generation—making it a genetic but non-hereditary feature.

Well, there seem to be heritable changes in gene activity which are not caused by changes in the DNA sequence, but by mechanisms such as DNA methylation and histone modification, each of which alters how genes are expressed without altering the underlying DNA sequence of the genome. They can turn an expressive gene off, resulting in the inability of genetic information to be read from DNA—so removing the methyl tag could turn the gene back on. The study of such phenomena is called epigenetics. It is the study of heritable changes in gene activity which are not genetic—that is, not caused by changes in the genotype. It is sometimes grouped under the banner of "soft inheritance"—a term coined by Ernst Mayr, referring to the inheritance of variations that are the result of non-genetic effects. Epigenetics is the field that seeks to explain features, characters, and developmental mechanisms in terms of interactions above the level of the gene.

How do such epigenetic processes work? This question calls in fact for a reductionist analysis of a holistic concept. DNA methylation is a biochemical process whereby a methyl group is added to the cytosine or adenine DNA nucleotides. DNA methylation may affect the transcription of genes in two ways. First, the methylation of DNA itself may physically impede the binding of transcriptional proteins to the gene, and second, and likely more important, methylated DNA may be bound by proteins known as methyl-CpG-binding domain proteins

(MBDs). DNA methylation can lastingly alter the expression of genes.

Second, there are histone modifications which can have a similar effect. They act in diverse biological processes such as gene regulation, DNA repair, chromosome condensation (mitosis) and spermatogenesis (meiosis). They can either activate or repress genes. Histones are highly alkaline proteins found in eukaryotic cell nuclei that package and order the DNA into structural units called nucleosomes. This enables the compression necessary to fit the large genomes of eukaryotes inside cell nuclei; the compacted molecule is 40,000 times shorter than an unpacked molecule. Histones are the chief protein components of chromatin, acting as spools around which DNA winds, and as such they can play an important role in gene regulation.

Another indication that "nurture" may be able to change "nature" is the following. A retro-virus is an RNA virus that is duplicated in a host cell using the reverse transcriptase enzyme which produces DNA from its RNA genome. This is actually a case of transcription in the "opposite" direction—the transcription of RNA from DNA in reverse, so to speak. Even human cells have this capability. Self-replicating stretches in the human genome known as retro-transposons utilize reverse transcriptase to move from one position in the genome to another via an RNA intermediate. Telomerase is another reverse transcriptase found in humans, which carries its own RNA template; this RNA is used as a template for DNA replication. Because this idea goes against an old dogma of genetics—a one-directional flow of genetic information—it may not have received as much attention in genetic research as it deserves. Not only does it go against an old dogma, but also against an old tradition of reductionism that reduces all of genetics to the level of genes and genotypes.

The tide seems to be turning, though. It is becoming more commonplace in genetics to accept that complex human traits arise from numerous genetic and environmental factors interacting with each other. Reacting to the casual way in which some scientists speak of a "gene" for depression or a "gene" for violence, one could argue that depression and aggression are only labels for rather complicated and variable patterns of behavior caused by various levels, including levels above the genetic level. The hypothetical genes some have come up with were once claimed, and then often had to be retracted; they were inventions that did not lead to discoveries. They are often based on the mantra of rigid reductionism, causing hypothetical genes to come and go—that is, until we allow for a more holistic approach.

For further reading:

Ideker, T., T. Galitski, and L. Hood. "A new approach to decoding life: systems biology." *Annu. Rev. Genomics Hum. Genet.* 2001, 2:343-372.

Kauffman, S. *The Origins of Order: Self Organization and Selection in Evolution.* Oxford University Press, 1993.

Mayr, Ernst. What makes biology unique? Considerations on the autonomy of a scientific discipline. New York: Cambridge University Press, 2004.

Verschuuren, Gerard M. *It's All in the Genes!—Really?* Charlestown, SC: Createspace, 2014.

11. Evolutionary Biology

The aim of Darwin's theory of natural selection is to explain the development and evolution of biological phenomena in terms of *adaptation*. Evolution is supposed to produce better adapted, more functional phenotypes. If trait X has a function Y, then Y is called "adaptive"—which means it has been successful in promoting survival and reproduction. As a consequence, X will become an adaptation, because any trait having a function increases the likelihood that organisms with that trait will survive longer and/or reproduce more in comparison with organisms who have less adapted traits. Consequently, a trait is not an adaptation in itself; it is rather an adaptation in comparison with other traits in a particular environment and with respect to particular criteria.

There may be some confusion regarding the terminology used here. Adaptation is usually understood as a process. However, as a practical term, adaptation is often used for the product itself—those features of an organism which result from an adaptation process. As a matter of fact, most aspects of an animal or plant can be correctly called adaptations, although there are always some features whose function is in doubt. It might be wise, though, to use the term adaptation for the evolutionary process, and the term adaptive trait for the bodily part or function that is the result of this process. In addition, adaptation may refer to both the current state of being adapted and to the dynamic evolutionary process that leads to the adaptation.

Soon Darwin discovered that structure and function may not always be in agreement. A trait that used to be an adaptive feature and had gone through a process of adaptation—such as the wings of ostriches and penguins that do no longer fly, or the webbed feet of upland ducks which rarely go near the water—may still be inherited while no longer beneficial. Environments happen to be volatile. Only an ideal world of a stable, unchanging environment would allow metazoans to almost perfectly adapt to the environment and establish a quasi-permanent equilibrium with it. However, our planet, far from a static entity, is a dynamical, ever-changing system (see chapter 12).

Besides, there are plenty of reasons why certain adaptive features may have been compromised—reasons such as genetic and developmental constraints, past history, opposing selection pressures (sexual selection versus predator selection), and many other factors. Constraints like these may keep adaptations from being more perfect than they actually are; they may be optimal—the best we have—but not perfect—the best there is. In addition, it is highly questionable whether all of evolution can be described in terms of adaptation. To just mention one single issue: Is the four-legged structure of vertebrates a better adaptation than the six-legged one of insects? It is hard to see how that could be true.

Because of all these considerations, it might be safer to replace the term "adaptation" with the term "fitness"—also called "Darwinian fitness" (to distinguish it from "physical fitness"). Functions can be measured in terms of fitness—being the expected or actual genetic contribution to future generations, which we called earlier D-fitness. Then there is also F-fitness defined as the ratio between the numbers of individuals with a certain trait after selection to those before selection. So it is likely that adaptations do contribute to the fitness of individuals in either way.

Charles Darwin has been praised by many for starting a reductionist approach in evolutionary biology by working with random variation in combination with selective retention. His keyword was the concept of natural selection in order to explain the diversity and advancement of living forms in nature. However, it has been a point of debate whether Darwin's approach is fully reductionist. There seem to be several holistic elements in his theory—even in its later version, Neo-Darwinism.

First of all, it could be argued that the very concept of natural selection resides on a higher level. A mutation, for instance, can be analyzed at a physico-chemical level, but the evolutionary effect of the mutation only makes sense within the larger context of the organism's adaptation to a specific environment. The laws of natural selection simply cannot be reduced to the laws of physics and chemistry, because there is no such thing as reproduction and selection in the physical sciences. Darwin's original model, by showing how organism-environment relations generate increased adaptation to new environments, is basically a form of interactionism under environmental constraints; if so, says Marjorie Grene, "it exhibits hierarchical, not reductive thinking."

Yet, some argue that Neo-Darwinism denies any kind of higher-level structure because it reduces its theories to their lower-level conditions. This discussion might indicate that reductionist and non-reductionist approaches exist alongside each other. Whatever stand we take, we should not forget that Darwin drew on the work of breeders, not of physicists and chemists, as a source for natural selection. This might be another indication that natural selection works at a higher level of organization.

Another holistic element in (neo-) Darwinism might be the use of the term *adaptation*. Adaptation is not even the bare outcome of natural selection, as selection only has the effect of "choosing" between options that are already available. Adaptation as a process is a higher-level concept that weighs traits of organisms in terms of their *design*. Only functional designs are successful in survival and reproduction. However, natural selection may explain that a fine working design has a better chance of being passed on to the next generation, but ultimately it cannot explain why such a design is working at all, let alone working so well. The fittest are not defined by their survival—that would create a tautology—but by their design. In artificial selection, the design is defined before breeding begins, but not so in natural selection.

That is where *teleology* keeps creeping in, given the simple fact that the term "design" is actually a teleological concept. Teleology embodies the idea that nature has "goals" and does things "in order to" or "for the sake of." Because the term teleology may carry some historical baggage—with connotations of intention, purpose, and foresight—biologists like to avoid the term in favor of the term *teleonomy*. This new term was introduced to describe the study of goal-directed functions which are not guided by the conscious forethought of man or any supernatural entity. In this sense, adaptation hoards hindsight rather than foresight. But it is still goal-directed. Criteria of fitness independent of survival are intricately connected with the *design* concept. Natural selection does not create the fit—it only selects what is a better fit.

Each time biologists are speaking of being "fit," "successful," or "goal-oriented," they are actually talking teleology—or teleonomy, if you wish. Certain biological features of organisms are "successful" and "effective" in reaching their "goal" because they have a design that has such a goal as an effect. If these biological features were not design-like, they simply would not work. Good designs must have "something" that carries them through the filter of natural selection. It has been rightly stated by Leon Kass that organisms are "not teleological or teleonomic because they have survived; on the contrary, they have survived (in part) because they are teleological or teleonomic."

In other words, the mechanism of natural selection does not explain but *assume* success and goal-directedness. As Michael A. Simon and Lynn Margulis have put it, "natural selection is only the editor and not the author of evolutionary change." It is a process that eliminates what is unfit, but does not create fitness. Therefore, Darwin did not and could not discard the concept of design—or what comes with it, goal-directedness. This made Michael Ruse exclaim that after Darwin, the heart still existed "for" circulation; the cause of its existence, he added, may have been different, but its goal-directedness was not. Speaking of "fitness" and "success" is not possible if one rejects goal-directedness.

Another issue of debate is what the basic units of evolution are—which is in essence an issue of ontological reduction. Some say it is genes (we discussed this viewpoint already in chapter 10). Others say it is organisms in their entirety, and not just their individual genes. Indeed, in terms of natural selection and adaptation, the individual seems to be the fundamental unit of operation. But when it comes to evolution, not even the individual qualifies, but rather the population or species. Organisms do not evolve, but populations and species do.

The task of evolutionary biology is to explain new species as arising out of earlier ones in the same way that mountains and lakes are explained as having arisen out of earlier geological formations. One could argue that only thinking in terms of populations, or even species—rather than individuals—allows us to formulate evolutionary theory. Since the members of any population or species differ from one another in various ways, and since there is no feature or set of features that is both necessary and sufficient for membership, there can be no individual or specimen to define membership. That which ties the members together is a genetic relationship. Therefore, it is not the individual but the population or species that is the basic unit of evolution (more on species later in this chapter).

Whatever stand we take in this discussion, there is a complex of abiotic and biotic factors in the organism's environment that may have an impact on selection. These factors are not only located at the level of organisms, but also at the level of populations. Apart from the mating partner, the composition of the entire population is of relevance to the reproduction rate of each member. What all these factors have in common is the fact that they affect the reproduction rate of organisms—which might indicate the organism's central position as a unit of selection. But even if so, the model of an organism should not be thought of as too simple; an organism has an intricate internal structure as well as an intricate web of connections with other organisms and its environment.

Because of this, the concept of natural selection should not be taken as a merely passive process. The widespread image of organisms adapting to their environment by natural selection neglects the fact that organisms also have the capacity to actively shape or even construct their environment by physical altering their surroundings, by selecting their environment, and by determining which parts of their physical surroundings constitute their environment. It is a two-way street. Organisms neither evolve in isolation nor simply adapt to "external" factors. On the contrary, they react to and modify their environment according to their own needs. Both organism and environment therefore undergo a continuous process of mutual interaction, which Richard Lewontin termed co-adaptation. As Fulvio Mazzocchi puts it, "A biological system is constrained by its environment, but it also changes its environment."

Yet, even those who keep maintaining that neo-Darwinism is typically reductionist in its approach should still ask themselves whether it can do an adequate job in explaining evolution as it has actually unfolded on Planet Earth. When looking at the fossil record, we find it to be full of gaps—forms appear and disappear suddenly. There are hardly any transitional forms. The gaps between fishes and amphibians (in spite of *Ichthyostega*), between amphibians and reptiles (in spite of *Seymouria*), and between reptiles and birds (in spite of *Archaeopterix*), to name just a few, are hard to conceive of as continuous transitions that were adaptations based on numerous small mutations of genes exposed to natural selection.

Yet, in Neo-Darwinism, organisms are considered "adaptation machines," which idea is supposed to be sufficient to generate evolution. But the question is whether all forms in the natural world are really a function of adaptation. Is adaptation really a universal and exclusive process in the living world? As a matter of fact, not all biological forms seem to be clear products of adaptation. Evolution may have produced endless variation, but on a rather limited set of basic designs.

Let us phrase the question again in this simple way: Why do vertebrates have a four-legged structure? Is that better than the six legs of insects? Once present, a four-legged structure could indeed evolve in specialized, adaptive ways for running, digging, jumping, and so on. But how did that structure itself come along? It doesn't seem very likely that the four-legged structure of vertebrates is more of an adaptation than the six-legged one of insects. If adaptation were all that counts in evolution, whales would be fish.

To prevent this implication, there must be some structural, or organizational, reference implied along with the adaptive one. Adaptation, pure and simple, is an insufficient instrument for the separation of different types of animals or plants, according to some biologists such as the paleontologist Otto Schindewolf. Those who agree with this objection tend to say that the sudden origin of new patterns of organization implies the recognition of order, as distinct from the statistical manipulation of the conditions producing order. This would be another call for a more holistic approach.

How does evolutionary theory deal with such issues that come more specifically from paleontology, morphology, and taxonomy? Should morphology be ahead of phylogeny, or rather the other way around? Darwinists would go for the primacy of adaptations at a lower level—small adaptations gradually adding up to big structures—whereas taxonomists would stress the primacy of structures at a higher level—so that future adaptations would be dictated by morphology, not the other way around.

Let's compare both views—the duality of continuous and small-scale conditions versus discrete and comprehensive patterns. In line with Marjorie Grene, we could label the two views as the Darwinists' view versus the taxonomists' view. The Darwinists' view claims that speciation is a universal, continuous, and gradual change of genotypes—gene-by-gene, step-by-step, mutation-by-mutation—although it does allow for slow changes in large, central populations as well as rapid speciation in small, peripheral isolates. The taxonomists' view, on the other hand, is focused on an underlying structural basis dictating the range of modifications that natural selection can work on; this would entail there are different degrees of trans-specific steps, depending on which structural border has been crossed.

Most paleontologists, morphologists, and taxonomists—unlike most other biologists—see the diversity of organisms in evolution through the lens of taxa that differ from each other due to their specific construction and configuration—which are organizational concepts at a higher level. According to this view, taxa are separated from each other by organizational borders, whereas Darwinists only acknowledge reproductive borders between species. But it is far from clear, according to taxonomists, whether organizational differences can be reduced to an accumulation of changes at the level of genes. It could very well be the other way around, making the organizational structure come first by determining which modifications are possible for natural selection to work on—which makes for quite a different, actually a more holistic approach.

If we take the term *genome* as representing more than a collection of expressive genes, but rather an entire *program* as to how and when to process these genes, we may need more than "gene hunting." A genome is more than a bunch of genes—it exists at a higher level (see chapter 10). Processes such as gastrulation and segmentation, for instance, may require more than a set of genes of the type that causes albinism and the like. A program also requires "tools" to activate, repress, and coordinate regular gene actions. When it comes to regular, expressive genes, chimpanzees and humans share about 98% of their genes, but that does not mean their programs work the same way. Programs operate at a higher level than that of genes. As we said earlier, a program does not do what a regular, expressive gene dictates, but it is the other way around—the expressive gene does what a program dictates.

The net result of this is that it may be possible for the operation of the genome to change in discrete steps, rather than incrementally. If we imagine a mutation affecting the switches of a regulatory pathway, then we have a mechanism that would explain discontinuous variation—without any disruption to the functional integrity of expressive genes. This could lead to a new type of organism, ultimately leading to a new species, or even higher taxa. Changes in the regulatory part may create different configurations and constructions—in other words, a different "build" in holistic terms.

The concepts of type, build, or organization in some form seems to be indispensable in this view. New forms of life embody new operational principles. Such results may not be as gradual and continuous as Neo-Darwinists expect them to be, but rather sweeping and drastic—"punctuated" rather than gradual, if you will. True, mutations in expressive genes may cause populations to change and even cross the species border (after geographical isolation), in which case the gene-pool model can be a fruitful tool. But, on the other hand, it is hard to imagine that these small modifications or adaptations could cross higher taxon borders, unless we accept changes at a higher level of the program or genome—in its regulatory part, that is.

Since some parts of the program reside at a higher level of control and regulation, it might be useful to distinguish different "thresholds" as an indication for the resistance against changes in the genotype. The species border displays a comparatively low threshold, low enough to permit further adaptation to the environment, and high enough to assure internal reproduction. On the other hand, there may be higher thresholds that coincide more or less with the borders created by taxonomists around an order, a class, or a phylum. So mutations may affect more than the regular parts of the genotype—namely, the switches and constraints located in its regulatory part. In the image we used earlier, natural selection is only the editor, not the author, of evolutionary change.

Are there any known genetic mechanisms for this kind of genetic changes to take place in the regulatory part of the genotype? Perhaps. Nowadays, there is growing evidence for "gene swapping" between species, also called horizontal or lateral gene transfer. The discovery of "jumping genes" (transposons) may make for another relevant mechanism. Let us not further discuss here the merits and shortcomings of these mechanisms in relationship to structural changes of the phenotype. But in general, one could say that this kind of mechanisms may have easily eluded us so far because they do not receive much attention from scientists with a reductionist outlook.

Regardless of whether we take the Darwinists' view or the taxonomists' view, evolution has created an impressive "tree of life," an image describing the relationship of all life on Earth in an evolutionary context. Most biologists will see this as a process of building up from the bottom up—which is a form of ontological reduction. The cosmos was presumably at one point a swirling mass of gas and dust out of which has come the extraordinary complexity of life as we experience it now. Simpler organisms evolved into more complex ones, and eventually sensation and thought made their appearance. It may seem that science is telling us that the arrow always goes from lower to higher, from simpler to more sophisticated, from chaos to order, from matter to form, from body to mind—with mind only emerging at the very end, if at all (see chapter 13).

This may easily give us the impression that everything in evolution is built (or builds itself) from the bottom up; in other words, that the most basic level of reality is the ontologically simplest and most trivial, and that everything emerges somehow out of the simplest elements. The root of the word "evolution" actually means "to unroll." However, this simple view might be too simplistic. The "slime" from which life supposedly arose was made of atoms that follow the elaborate laws of electromagnetism and quantum mechanics, which in turn come from the even more profound structures studied in "quantum field theory" (see chapter 7). As one moves deeper into nature—to lower levels about which biologists can tell us nothing—one encounters, in the words of Stephen Barr, "not less and less form but increasingly magnificent mathematical structures, structures so profound that even the greatest mathematicians are having difficulty understanding them."

Evolutionary biology also has to deal with the question of how "real" certain evolutionary entities are, especially when it comes to what we call a "species." We saw earlier that a species is seen by many biologists as one of the levels in the organizational hierarchy of life (see chapter 2). In this view, a species is a unit of organisms, just as an organism is a unit of cells. This species level is called a species *taxon*. However, to make things more confusing, the species level is also a certain "level" in the Linnaean hierarchy of taxonomy. This level makes for a species *category*. In the species category, we find groups which taxonomists assemble into categories at a still higher level in the Linnaean hierarchy (such as genus, family, order, phylum).

Species taxa puzzle evolutionary biologists as to how they came into being. The difficulty of understanding such a slow historical process has repeatedly driven scientists out of their laboratories and into the arms of philosophy. Therefore, it may be worthwhile to dwell here briefly on one of the more recent issues in the reductionism-holism debate: What is the status of a biological species?

There is an important philosophical distinction to be made between an individual and a class. A molecule, a cell, and an organism are considered to be individuals; individuals can be identified in space and time, and may consist of parts. A class, on the other hand, may contain members but never parts; its members can be *identified* in space and time, but this is not true of the class itself. The class is *defined* with membership determined on the basis of certain criteria.

What, then, is a species? Is it an individual or a class? There is no doubt that a species category is a class—a level in the Linnaean hierarchy of taxonomy. But what about a species taxon—a level in the hierarchy of life? The species taxon *Homo sapiens*, for instance, is a member of the species category and belongs to the class of species taxa. But is this concrete group of organisms in itself also a class? Put in more general terms: Is a group of organisms to which the rank of species is assigned also a class—or is it rather an "individual," a real entity?

Some biologists, of the reductionist type, consider a species taxon to be a mere class—just like a species category is one. They treat a particular species of organisms in the way chemists treat a particular "species" of molecules—namely a class of members defined in terms of properties essential for membership in the class. In this view, the species *Homo sapiens*, for instance, is supposed to be a class of similar organisms. As a consequence, the name "Homo sapiens" would be like a kind name, not a proper name. In this view, a species in the life sciences would not consist of parts, but of members which are clustered together according to certain criteria. Membership in a class is then strictly determined on the basis of similarity; members can be identified by their defining properties. The consequence of this view would be that criteria common to all members have to be found, which entails a listing of traits or similarities such as a common morphology, exemplified by the so-called holotype specimen. This approach leads to a typological species concept. It is the degree of differences that is used as a criterion by which it is decided whether certain individuals belong to the same species taxon or not.

What is the problem with this concept of species? First of all, it introduces some arbitrariness into deciding how different a population has to be to deserve species status. Second, species taxa cannot have constant diagnostic characteristics because species constantly evolve. Third, there are many so-called sibling species which are indistinguishable on the basis of their appearance and yet are separate units which do not interbreed with each other. And fourth, we may end up with a static species concept, which is based on an underlying, ideal type shared by all the members of the same species taxon at a certain stage of evolution. It is clear that the typological species concept does not fit well in a neo-Darwinian theory of descent. It entails, in essence, a reductionist version of a species claiming that there is no such "thing" as a species, but only individual organisms. If so, a species is only a fictional product of classification.

For this reason, most evolutionary biologists prefer a more holistic approach by treating a species taxon as an individual instead of a class—that is to say, an individual that can (in principle) be identified in space and time, and may consist of parts. In this view, a species taxon would be an "individual" similar to the way an organism is. As a result, a species taxon of the species category would be a concrete zoological or botanical object, such as the "whole" of all wolves or the "whole" of all red oaks. They are not members of a class but parts of a whole. This holistic approach makes a species taxon a spatial and temporal entity—that is to say, the most fundamental unit in evolution.

So then the question arises: What makes a species an identifiable unit? What keeps a species taxon together and what distinguishes it from other species taxa? A clear answer to this question has been given by a species concept that has proven rather successful among biologists, the so-called biological (or reproductive) species concept. It was particularly advocated by Theodosius Dobzhansky and Ernst Mayr and has been phrased as following: A species is a population of organisms that can breed with one another, but not with other populations. The members or parts of a species taxon, according to the biological species concept, are bound together by gene flow.

According to this concept of a species, a species taxon is an evolutionary unit kept together by gene flow and isolated from other taxa by intrinsic isolating mechanisms. Its members—or rather its parts—may share some similarities, but what is more important is the fact that they are internally connected by gene flow but isolated from other groups by reproductive barriers. What is essential here are family ties, not family resemblances. Similarities as such do not count for much, as they are believed to be a mere by-product of reproductive isolation; theoretically, the parts of a species may not even have anything in common but reproductive isolation.

The strength of the biological species concept is that it relates our understanding of the concept of a species to our understanding of speciation processes. It offers an explanation of what maintains and disrupts the unity of the species taxon. Although there may be different degrees of unity within a species taxon, the core message of this holistic approach is that a species can be seen as a real entity at its own organization level with a whole-part relationship.

For further reading:

Ereshefsky, M. ed. *Units of Evolution: Essays on the Nature of Species.* Cambridge, MA: MIT Press, 1992.

Grene, Marjorie. *The Understanding of Nature: Essays in the Philosophy of Biology.* Boston, MA: Reidel, 1974.

Kass, Leon R. *Teleology, Darwinism and the Place of Man: Beyond Chance and Necessity?* New York: Free Press 1988, chapter 10.

Ruse, Michael. *Philosophy of Biology.* Albany, NY, 1988.

Verschuuren, Gerard M. *Darwin's Philosophical Legacy—The Good and the Not-So-Good.* Lanham, MD: Lexington Books, 2012.

12.   Ecology

The nested organizational hierarchy we have used so far—ranging in scale from atoms, to molecules, to cells, to tissues, to organs, to organisms—can even be extended further: to populations, to species, to communities, to ecosystems, and up to the level of the biosphere. This extended hierarchy invites ecologists to work at a variety of levels within the field of ecology itself, either in a reductionist way or a holistic way. The difference between these two approaches may be a relative one. A theory about communities may be holistic with respect to some lower level of organization, such as populations, but at the same time reductionist with respect to some higher level, such as ecosystems. The wholes on one level become parts on a higher level.

In its epistemological version (see chapter 5), the reduction problem in ecology has been very elusive. There has hardly been any successful reduction of concepts or theories in ecology to lower-level concepts and theories. For this reason, holism-reductionism disputes in ecology are primarily concerned with ontological and methodological issues, which are strongly intertwined. As Rick C. Looijen puts it, "Ecologists are constantly discussing what should be the 'fundamental' unit of ecological research and what research strategy should therefore be followed." Population ecology deals with groups of organisms of the same species, but community ecology with the interrelationships between groups of organisms of different species of plants and animals. And systems ecology deals with the complex of relations between various groups of organisms and the abiotic environment, which together make up an ecosystem.

Reductionist ecologists are primarily population ecologists; they take modern evolutionary theory as a starting point, and argue that populations should be the fundamental units of ecological research. Holistic ecologists, on the other hand, are primarily systems ecologists; they argue that the "functional" level of organization in ecology is formed by ecosystems, and that research should therefore be directed at these systems. An ecosystem is generally defined as the whole of a biotic community—also called a biocoenosis—of plants, animals, micro-organisms, and its abiotic environment.

Apparently, this dispute has everything to do with the ontological status of communities and ecosystems, which is actually the core of the reduction problem in ecology: What is the status of communities and ecosystems? Are they "real," discrete ontological entities with emergent properties (such as biomass, productivity, diversity, stability), which are assumed to be largely independent of the constituent species and populations? Or are they nothing but accidental assemblages of species and populations which just happen to share the same space and time, so they can be explained fully in terms of individual adaptations of the species and populations involved? Of course, the answers to this question largely determine the methodological positions taken. Looijen summarizes this as follows, "Those who assume that communities and ecosystems are 'real' ontological entities argue that research should be directed at these higher levels of organization, whereas those who deny their existence naturally argue that research at these levels is pointless."

Of course, there are many intermediary and mixed positions in this debate. A common position is that, on the one hand, one recognizes that communities and ecosystems are real ontological entities while, on the other hand, it is assumed that their structure and dynamics are determined primarily by interactions between species. In this latter view, research should logically be directed at both higher and lower levels. This is a moderate or mixed approach in which it is acknowledged that communities and ecosystems are discrete higher-level entities having their own emergent properties, but it seeks the explanation of these properties especially in interactions between the component species. In the words of the legendary ecologist Eugene Odum, "New properties emerge because the components interact, not because the basic nature of the components is changed."

In what follows, we will focus mainly on the holistic approach of ecology, as it is the most controversial and yet most characteristic one. It claims that the entities of higher levels of organization—populations, communities, ecosystems, biosphere—are discrete integrated wholes having their own emergent structure and dynamics, which cannot possibly be derived from properties of the component elements. Characteristic of this view is that populations, communities, and ecosystems are regarded as analogues to organisms—some sort of "super-organisms." What does this mean?

Organisms are usually seen as "masters in self-regulation." They control and steer processes within the organism so as to correct for accidental changes surrounding them, which is called homeostasis. Organisms do so, among other things, with negative feedback loops that dampen any oscillations of internal variables. This phenomenon is collectively called self-regulation. Blood, for instance, is constantly being checked for its temperature, pressure, acidity, and so on. This way, such factors oscillate only within a narrow margin around a specific norm maintained by some kind of "thermostat." When, for instance, the blood glucose level increases, the concentration of insulin goes up, which makes the concentration of glucose go down again. A phenomenon like this can certainly be analyzed at lower organizational levels, but it also is a candidate for a more holistic approach, because it requires higher-level norms regulated by built-in "thermostats" and feedback loops.

It is very tempting, especially from a holistic point of view, to assume something similar at levels *above* the level of organisms. Can entities on those higher levels be seen as some kind of "super-organism"? Do populations, species, communities, and ecosystems show forms of "self-regulation" similar to what organisms do—which some ecologists consequently call a "natural equilibrium"? Let us discuss such holistic claims for populations, communities, ecosystems, and biosphere separately.

1. Populations, seen from a holistic perspective, are self-regulating entities which occur in equilibrium with their environment. They are believed to exhibit homeostasis and other organismal properties which would control the growth of populations. Contrary to those who seek to explain population numbers in terms of extrinsic factors, the self-regulatory school concentrates on intrinsic factors within populations, such as individual differences in behavior and physiology. An outspoken representative of this school was V. C. Wynne-Edwards (1962) who became particularly famous for his ideas on group selection.

2. Communities, seen from a holistic perspective, are also "super-organisms" which, analogous to real organisms, have a certain physiological integrity and go through a process of coming into existence, growing, and then developing into a stable "mature" stage. The standard example of this is the view of the American plant ecologist Frederic Clements whose theory was primarily meant to explain a phenomenon in the plant world known as (primary) succession. This is the development of a plant community starting with the first plant growth on a formerly bare soil, and then progressing in the direction of a more or less stable end phase, the so-called climax stage. According to Clements, this succession cannot be explained in terms of adaptations of each individual species to local environmental conditions, but is mainly the result of interactions between species, in particular through competition with each other.

3. Ecosystems are, from a holistic point of view, also seen as "super-organisms." According to the best known representative of this school, Eugene Odum, ecosystems would have a "strategy" showing a development (succession) from a young to a ripe or mature climax stage, which is an end stage considered to be stable. The norm of the system, "stability," would be achieved and maintained in ecosystems through various feedback mechanisms and an information network in the form of food webs and nutrient cycles. It is assumed that there is a relation between the diversity of a system and its stability: The more diverse an ecosystem is, the more stable it will be. Yet in this view, it is not the organisms or populations that form the units of ecosystems research, but so-called trophic levels (or food levels)—namely, primary producers (green plants), first order consumers (plant eaters), second order consumers (meat eaters), and reducers (fungi and the like). In the words of Looijen, "These are conceptual black boxes, or theoretical compartments, through which energy and matter flow on their way through the system."

4. Biosphere would be the highest level in the ecological hierarchy. It is the global sum of all ecosystems integrating all living beings and their relationships, including their interaction with the elements of the lithosphere, hydrosphere, and atmosphere. During the early 1970s, the British chemist James Lovelock developed the so-called *Gaia* hypothesis

which claimed strong ties between the biosphere and other Earth systems. For example, when carbon dioxide amounts increase in the atmosphere, plants grow more quickly; as their growth continues, they remove more and more carbon dioxide from the atmosphere. The idea behind this hypothesis is that the biosphere is again taken as some kind of "super-organism" that exhibits features of self-regulation and homeostasis based on built-in norms—just like regular organisms usually do. As a consequence, the biosphere will maintain its homeostasis against any disruptions such as the ones caused by the greenhouse effect, for instance.

What all these holistic views in ecology have in common is the characteristic of self-regulation. This would supposedly be achieved through various feedback mechanisms and an information network similar to the physiological mechanisms that maintain homeostasis in organisms. We discussed earlier (in chapter 2) that at each organizational level in the life hierarchy properties may be found which fail to occur at other levels in the hierarchy. Perhaps the property of self-regulation is one of them. It certainly works at the levels of organisms, for they are usually good at self-regulation thanks to different kinds of "thermostats."

The question remains, though, whether the concept of self-regulation can also be applied to ecological levels. Obviously, there are some similarities. Take, for instance, the case of predator-prey relationships as portrayed in the classical competition models of Lotka and Volterra. They seem to follow the following regulation pattern: When the number of predators rises, the number of prey animals goes down, which makes the number of predators go down also. At first sight, this looks very much like the self-regulation, or equilibrium, of a "super-organism," but the question is whether there is really a norm involved here, and how it was set.

Some ecologists have suggested that perhaps the "carrying capacity" of the environment is the norm searched for. However, this concept is not very helpful here as it refers to a maximum, not a norm. Is the norm perhaps equivalent then to something like the average number of predators and prey animals? But averages are not norms either; they tend to fluctuate due to changes in the environment, whereas self-regulation is supposed to stabilize them. Besides, there is the problem that the norms behind self-regulation in organisms are the outcome of forces such as natural selection. But how would natural selection work on entities such as ecosystems, not to mention the biosphere as a whole?

The most common response among reductionists in ecology is that terms such as homeostasis, equilibrium, and self-regulation suggest something at a higher level that is not really there. Therefore, they suggest that the concept of "stability" is a much safer term, because it is more of a statistical nature and refers to large numbers of lower-level factors; these factors are independent of each other, but due to their multiplicity, can restrain each other or level each other out—without actually regulating each other. A "stability" like this is rather seen as a matter of statistical risk-spreading at a lower level than of regulation at higher levels. This approach would also explain that the more species diversity a system has, the more stable it can be. A tropical rainforest, for instance, has more "stability" than an agricultural field. This does not mean, of course, that small ripples in a rainforest can be so numerous and widespread that they still affect the "stability" of the system. The magnitude and the duration of oscillations are two different issues.

All these considerations seem to make a case for a more reductionist approach in ecology. If there is stability at a certain ecological level, the explanation may not be self-regulation but rather a form of risk-spreading based on large numbers of interacting factors. Because this approach acknowledges interactions between lower-level entities, it is not reductionist in the strict sense. Yet, randomness has taken over the position of some higher-level concepts. The general idea behind this reductionist view is, according to Looijen, "that in searching for explanations of community structure, one should first assume that this is purely the result of chance processes before looking at deterministic, biological explanations." Therefore, Looijen recommends, "Explanations of the so-called 'structure' of communities should perhaps always be tested against the null hypothesis that this structure is completely random, given the properties of individual species and the environment."

Even the distribution and number of species within an ecosystem can be seen in terms of randomness, for they are highly dependent on immigration and environmental conditions, which are both subject to randomness. Communities are assemblages of species that gradually flow into one another through migration. As Looijen remarks, "Environmental conditions are rarely, if ever, discretely distributed but vary continuously, both in space and in time." If environmental conditions do indeed vary continuously over time, there could not even be a deterministic succession towards one stable end phase, as holistic ecologists would claim. On the other hand, holism is not as deterministic as reductionism is because it allows for a certain degree of unpredictability in complex systems.

If these considerations are right, randomness might also overhaul the assumption that communities and ecosystems are clearly distinct entities on their own. If biotic communities, or their sub-communities of plants and animals, are not discrete entities, then neither can be ecosystems. Some population ecologists therefore argue that communities and ecosystems do not exist at all, at least not as discrete, structurally and/or spatially distinct entities; instead they are seen as accidental collections of populations. So a "community" would merely be a name for a group of individuals, gathered together more or less at random, in a place where they find suitable environmental conditions, without hampering their mobility.

Let us wrap up our concise discussion of a complex issue. Strict holism, on the one hand, may provide ecology with new concepts that can stimulate research. However, they would still need thorough further testing. Strict reductionism, on the other hand, might end up in a mere description of numerous variables without much of a structure or organization. A more moderate combination of both approaches seems to have the best prospect of advancement in ecology by acknowledging interactions between components. Perhaps most ecologists can at least agree with what Martin Rees said about all higher-level sciences: If it is true that many problems in the environmental sciences remain unsolved, "it is because it is hard to elucidate their complexities—not because we don't understand subatomic physics well enough."

The important lesson might be that scientific understanding of particular phenomena such as predator-prey relationships does not necessarily involve elimination or replacement of high-level entities and processes (such as populations, energy cycles, and so on) with lower-level entities and processes (molecules, chemical reactions, etcetera).   Again, we do not seem to have an "either-or" issue here.

For further reading:

Levins, Richard, and Richard C. Lewontin. "Holism and Reductionism in Ecology." *Synthese* 43 (1980) 47-78.

Looijen, Rick C. Holism and Reductionism in Biology and Ecology: The Mutual Dependence of Higher and Lower Level Research Programmes. Kluwer Academic Publishers, 2000.

Wilson, David Sloan. "Holism and Reductionism in Evolutionary Ecology." *Oikos*, Vol. 53, Fasc. 2 (Sep., 1988), pp. 269-273.

13.   Neuroscience

Neuroscience has become the flagship of reductionism. The findings by contemporary neuroscientists that specific regions or scattered systems in the brain are associated with particular cognitive and emotional functions provide compelling evidence that the human mind has something to do with these capacities and their localizations. Scanning techniques such as an electro-encephalogram (EEG) or functional magnetic resonance imaging (fMRI)—an MRI procedure that measures brain activity by detecting associated changes in blood flow— have opened up the brain for us as the base of all mental activities.

Partly due to such discoveries, the biologist and neuroscientist Francis Crick ventured to say, "You're nothing but a pack of neurons." Back in 1994, he wrote, "The astonishing hypothesis is that you, your joys and your sorrows, your memories and your ambitions, your sense of personal identity and free will are, in fact, no more than the behavior of a vast assembly of nerve cells and their associated molecules." Marvin Minsky, a pioneer in the field of artificial intelligence would add to this that human beings are mere meat-machines—yes, machines made of meat. The list of similar reductionist claims could go on and on.

The reductionist approach of neuroscience has been heralded as very successful, at least in the popular press and in the general perception. The common view is that whatever happens in the brain, or whatever the human mind does, can be and has in fact been effectively analyzed in terms of lower-level elements such as nerve cells and their associated molecules such as neurotransmitters. Yet, some say in response, this "flat" picture of neuroscience does not do full justice to what is actually going on in neuro-scientific research. Most neuroscientists will rather characterize their own research as spanning several levels—although "levels talk" is not very common in this field.

When talking about levels, we have to keep in mind that there is hardly any agreement on the definition of a level, or on the criteria for distinguishing levels (see chapter 2). The levels we discussed so far—referring to atoms, molecules, cells, organisms, etc.—are very common in philosophy, but in neuroscience they are usually of a different nature. When the term *level* does appear in neurological literature or research, it is most often used in reference to various levels of processing, such as the different stages of visual information processing—spanning the retina, the LGN, the visual cortex, and so on. As the cognitive scientist Markus I. Eronen remarks, "How many levels there are, and which levels are included, are questions often answered on a case-by-case basis by discovering which components at which size scales are explanatorily relevant for a given phenomenon."

In general, it is assumed that in between the lowest and highest level in neuroscience reside various levels of neural systems, such as the frontal cortex, parietal cortex, hippocampus, and basal ganglia, plus the interactions among all these systems, and their modulations by neurotransmitters in response to relevant task events. The existence of optical illusions, for instance, calls for a multi-level approach; they show that humans perceive more than the sum total of the sensations on the retina.

A more specific example of this multi-level approach would be the hierarchy of motor actions. The neuroscientist James Knierim, for instance, describes the following levels. The spinal cord and brainstem are involved in the low-level "nuts and bolts" processing which controls the activity of individual muscles. Individual alpha motor neurons control the force exerted by a particular muscle, and spinal circuits can control sophisticated and complex behaviors such as walking and reflex actions. However, the types of movements controlled by these circuits are not initiated consciously. Voluntary movements require the participation of the third and fourth levels of the hierarchy: the motor cortex and the association cortex. These areas of the cerebral cortex plan voluntary actions, coordinate sequences of movements, make decisions about proper behavioral strategies and choices, evaluate the appropriateness of a particular action given the current behavioral or environmental context, and relay commands to the appropriate sets of lower motor neurons to execute the desired actions.

Another example that Eronen mentions would be the case of spatial memory and long term potentiation (LTP). At the top of the hierarchy, there is the level of spatial memory, which involves various types of memory and learning. The level of spatial map formation includes the structural and computational properties of various brain regions involved in spatial memory, most importantly the hippocampus. The cellular-electrophysiological level includes neurons that depolarize and fire, synapses that undergo LTP, action potentials that propagate, and so on. At the bottom of this hierarchy is the molecular level, where we find NMDA and AMPA receptors, Ca^{2+} and Mg^{2+} ions, etcetera. Entities at each lower level are components in a higher-level mechanism; the hippocampus, for example, is an active component in the spatial memory mechanism, synapses are active components in the hippocampal mechanism of memory consolidation, and finally, NMDA receptors are active components of the synaptic mechanism of LTP.

Apparently, exploring the complexity of the nervous system requires different analyses at multiple levels. As a result, neuroscientists distinguish various specializations in neuroscience. They investigate how genes and molecules regulate nerve cell function (cellular/molecular neuroscience), they explore how neural systems produce integrated behaviors (behavioral neuroscience), they seek to understand how neural substrates create mental processes and thought (cognitive neuroscience), and they study the impact of social factors (social neuroscience). Accordingly, neuroscience research works with models that span various levels as well, ranging from biophysically detailed models of ion channels, synaptic transmission, and dendritic integration, up through systems-level models of visual perception, sensory-motor control, memory, and language.

Is this reductionism? Yes, it is when activities at higher levels are considered to be completely determined and explained by activities at the lower, and ultimately at the lowest, level. But it is not full-fledged reductionism when the different levels are believed to make for different kinds of *top-down* causation. In this latter view, it is still possible that new, emergent properties appear at higher levels and may affect what happens at lower levels. Downward causation means that higher level entities are constraining conditions for the activities of lower levels. In the words of the Nobel Laureate Roger Sperry, "Even the brain cells, however, with their long fibers and impulse-conducting properties, do not have very much to say about when they are going to fire their messages, for example, or in what time pattern they will fire them. The firing orders for the day come from a higher command." Something similar can be said about phantom pains; any groans they may elicit are caused not by the biophysics, chemistry, or physiology of the cerebral nerve impulses as such, but by the pain property itself.

The most controversial part of the reductionism-holism debate regards the highest level of brain activities, commonly referred to as the mind. To illustrate what this debate is about, let us study the example of "split brain" research as done by Roger Sperry and by Michael Gazzaniga.

The "split brain" phenomenon occurs when the corpus callosum, connecting the two hemispheres of the brain, is severed to some degree. Communication and coordination between the two hemispheres is essential because the two hemispheres have some separate functions. The right hemisphere of the cortex excels at nonverbal and spatial tasks, whereas the left hemisphere is more dominant in verbal tasks, such as speaking and writing. The right hemisphere controls the primary sensory functions of the left side of the body. In a cognitive sense, the right hemisphere is responsible for recognition of objects and timing; in an emotional sense, it is responsible for empathy, humor, and depression. On the other hand, the left hemisphere controls the primary sensory functions of the right side of the body and is responsible for scientific, mathematical, and logical skills.

Tests were done on various "split brain" patients. When stimuli were flashed to the right visual field and then processed by the brain's left hemisphere, which contains the language center, patients were able to press a button to indicate they saw the stimulus and could verbally report what they had seen. However, when the stimuli were flashed to the left visual field, and thus processed by the right hemisphere, they would press the button, but could not verbally report having seen anything. When the experiment was modified to have the patients point to the stimulus that was presented to their left visual field and not have to verbally identify it, they were able to perform this task accurately.

The New York Times reported on October 31, 2011, "In the decades to follow, brain scientists found that the 'left-brain right-brain split' is only the most obvious division of labor; in fact, the brain contains a swarm of specialized modules, each performing a special skill such as calculating a distance or parsing a voice tone, and all of them running at the same time, thus communicating in widely distributed networks, often across hemispheres." So the pressing question is: Why, if we have these separate systems, is it that the brain still has a sense of unity? Where does the coherence of all these interactions come from? It is here that the reductionist approach and the holistic approach part. Let us analyze both interpretations.

Reductionists would seek to localize the unity of brain activities in some specific part of the brain. The brain's cacophony of competing voices feels coherent, they say, because some module or network somewhere in the left hemisphere is providing a running narration. Gazzaniga decided to call this left-brain narrating system "the interpreter." The interpreter creates the illusion of a meaningful script, as well as a coherent self. Consciousness is thus seen as a local phenomenon, supposedly happening in many places, but only the "interpreter" is conscious of the information that it receives. This view may not be extreme reductionism. Roger Sperry, for his part, does acknowledge that the higher order mental patterns and programs, once generated from neural events, have their own subjective qualities; they supposedly progress, operate, and interact by their own causal laws and principles which are different from and cannot be reduced to those of low-level neurophysiology. He even says, "The mental entities transcend the physiological just as the physiological transcends the [cellular], the molecular, the atomic and subatomic."

This latter position is sometimes called "biological naturalism." Its basic idea is that mental states, though not identical to the firing of neurons or any other brain processes, are nevertheless caused by such processes in a manner analogous to the way the solidity of an ice cube is caused by the state of the water molecules composing it. Consciousness and other mental phenomena are thus higher-level features of the brain, just as solidity is a higher-level feature of the system of water molecules constituting the ice cube. But just as solidity is nevertheless a physical property of a system of water molecules, so too is consciousness presumably a physical property of the system of neurons constituting the brain. Those who claim that mental concepts are merely products of neurons should realize that talking about neurons requires the concept of *neuron* to begin with.

We seem to be bordering here a more holistic approach. Holism does not only acknowledge multiple levels, giving psychological and mental events their own level, but it also claims that emergent properties at a higher level may have a downward causation effect on lower levels. Not only would neural processes cause mental processes, but also the other way around. One could even make the case that Sperry's concept of consciousness in fact acts as a causal agent at a higher level having downward influence over cerebral functions. This approach has caused a major shift in neuro-biological thinking.

However, this more rigorous holistic claim is still rather controversial among many neuroscientists. Are there really such things as psychological and mental events, and if so, can they interfere with lower levels? The holistic approach would affirm that they do exist, and then point out that certain *psychological* states (e.g., prolonged anxiety or embarrassment) can cause physiological effects (heightened blood pressure, eczema, blushing) in a human body. And certain *mental* states (e.g., anticipation, self-affirmation, phobias) can cause physiological effects as well (placebo effect, self-improvement).

The pivotal issue here is whether mental states such as beliefs and decisions do play a causal role in the production of behavior. Whereas reductionist neuroscientists would maintain that our beliefs, values, memories, choices, and aspirations are in fact *illusory*—not causal, that is—their holistic colleagues would emphasize that human actions are indeed caused, at least in part, by a person's beliefs, values, memories, choices, and aspirations. In the words of Roger Sperry, "Above simple pain and other elemental sensations in brain dynamics, we find, of course, the more complex but equally potent forces of perception, emotion, reason, belief, insight, judgment, and cognition. In the onward flow of conscious brain states, one state calling up the next, these are the kinds of dynamic entities that call the plays. It is exactly these encompassing mental forces that direct and govern the inner flow patterns of impulse traffic, including their physiological, electro-chemical, atomic, subatomic, and subnuclear details."

In this latter view, it is *cognitivity* that makes the difference. Cognitivity pertains to the mental processes of perception, memory, judgment, and reasoning. It is these encompassing mental forces that presumably direct and govern the inner impulse traffic, including its electrochemical and biophysical aspects through downward causation. The cognitive scientist Jerry Fodor put it most vividly and dramatically: "If it isn't literally true that my wanting is causally responsible for my reaching, and my itching is causally responsible for my scratching, and my believing is causally responsible for my saying... if none of that is literally true, then practically everything I believe about anything is false and it's the end of the world."

The idea of downward causation from the mental to the neural seems to be accepted, or even essential, in at least one particular field, the field of *cognitive* neuroscience. An instructive example of this is the recent finding that acute depression is more effectively treated by a combination of cognitive-behavioral therapy and antidepressant drugs than by either alone. Cognitive-behavioral therapy is typically a high-level intervention in which the explicit goal is to change the patient's belief structures and modes of interacting with the world. Pharmacological treatment, on the other hand, is a low-level intervention in which the explicit goal is to manipulate the synaptic concentration of the neurotransmitter serotonin. In a combined treatment, the causal potency of a belief, an idea, or an ideal becomes just as real as that of a molecule, a cell, or a nerve impulse.

If there is such a thing as downward causation, then it seems to follow that ideas can cause ideas and help evolve new ideas to "steer" the brain and the body. In the words of Roger Sperry again, "Near the apex of this compound command system in the brain we find ideas. In the brain model proposed here, the causal potency of an idea, or an ideal, becomes just as real as that of a molecule, a cell, or a nerve impulse. Ideas cause ideas and help evolve new ideas. They interact with each other and with other mental forces in the same brain, in neighboring brains, and in distant, foreign brains."

Somehow, we are entering here a much broader discussion, taking us inevitably from the domain of science into philosophical territory. It is tempting to rephrase the distinction between neural states and mental states as a distinction between brain and mind. Certain mental capacities such as intelligence are often considered a "mind" issue, prompting the question of how the mind could possibly "steer" the brain. Monica Anderson astutely remarks about a mental issue such as intelligence, "It must emerge from the interactions of non-intelligent components. In some sense, this statement is trivially true, since if we had intelligent components we'd be done before we started. But the design of low-level components like simulated neurons to generate emergent effects like intelligence is still largely unexplored territory." That's why the gap between brain and mind seems to remain acute.

No wonder, then, that the philosophical brain-mind distinction remains highly significant—not only in psychology but also in neuroscience. An interesting case in this context would be the *attention* issue. Attention is the cognitive process of selectively concentrating on a discrete aspect of information, while ignoring other perceivable information. When someone says, "This caught my eye," the person's attention is more or less passively triggered by some specific cue or stimulus. This seems to be mainly a *brain* issue. If a neuron has a certain response to a stimulus when the individual is not attending to the stimulus, then when the person does attend to the stimulus, the neuron's response will be heightened even if the physical characteristics of the stimulus remain the same. On the other hand, when someone says, "I did not pay attention," we seem to be dealing with a *mind* issue. It results from an active decision to pay, or not to pay, attention. Here we have the role of attention in the everyday sense of actively focusing on or taking interest in something, of voluntarily directing our attention toward some cognitive object. It is not a passive but an active response.

In addition, recent findings suggest that the mind's mental states are even able to install new "programs" in the brain by creating new neuronal connections. Neuroplasticity is the key term for the flexibility of brain structures, based on the now proven understanding that changes in behavior and environment actually alter the neuro-circuitry of the brain. It is currently common knowledge that the understanding of brain plasticity has replaced the formerly held position that the brain is a static and unchangeable organ. Arguably, mental states of the mind can act as a causal force in the brain, able to change the material brain's synaptic structures, by way of downward causation.

It is through goal-setting, self-talk (affirmations), self-examination (evaluations), and mental imagery (visualization) that the mind can create new connections in the brain. B. Forstmann and E. Wagenmakers tell us, "For example, when humans perform a 'reinforcement learning task,' they are not only incrementally learning probabilistic stimulus-action-outcome associations and choosing between them, but they also engage in other cognitive strategies involving hypothesis testing and working memory." Cognitivity is key here again.

To sum up, a comprehensive explanation of behavior usually requires multiple levels of analysis. Social-cognitive neuroscience, for instance, is an emerging scientific discipline that attempts to integrate the theories, methods, and insights of two rather new fields: social cognition and cognitive neuroscience. In the words of the neuroscientist M. Liebermann, "Although these disciplines share a focus on the information-processing mechanisms that underlie behavior, they move beyond those mechanisms in opposite directions." Whereas cognitive neuroscience moves "downward" into the brain—with the aim of relating particular mental abilities (such as visual-spatial attention, working memory, and so forth) to the structure and function of lower-level neural systems— social cognition, on the other hand, moves from the information-processing mechanism "upward," into the phenomenology of the person himself or herself, exploring the social, cognitive, and affective forces that motivate particular behaviors, and the consequences that follow from them (see chapter 15).

The central issue in our discussion so far has been whether *mental* states can be fully equated to *neural* states. Reductionists would say so. But most neuroscientists of the holistic approach would deny this equivalence and maintain that mental states not only reside at a higher level and can cause downward causation on the lower level of neural states, but they may be even of a different nature.

One of their arguments is that neural activity is not a *sufficient* condition for mental activity. Mental events are believed to transcend neural events to the same degree by which life transcends chemistry and physics. Even if we discover that certain mental phenomena are associated with certain neural phenomena, this does not entail that these mental phenomena were *caused* by neural phenomena—the causation might even go in the opposite direction. Correlation doesn't automatically equal causation. The fact that regions light up on an fMRI does not explain whether this lit-up state indicates they are causing a certain mental state, or just reflecting it.

Therefore, we need to find out whether certain mental phenomena always correlate with certain neural phenomena; and furthermore, whether they are proportional to the intensity of the mental phenomena. If such is not the case—and it is claimed by some that mounting evidence indicates it is not— then the reductionist claim of reducing the mental to the neural has actually been falsified. Whereas something like pain, for instance, can be induced in a physical way, there is arguably no evidence that experimental stimulation of specific neuronal areas could produce a specific mental state, let alone a specific thought. Even lie detector tests do not detect thoughts, but at best physiological and emotional responses to those thoughts.

Another argument for the distinction between neural and mental states is that neural activity not only fails to be a sufficient condition for mental activity, but it may not even be a *necessary* condition. Put differently, mental activity doesn't always correlate with neural activity; in fact, there may be mental activity without hardly any neural activity. The need for distinguishing mental events from neural events becomes even stronger when we consider situations where the most intense subjective experiences correlate with a dampening—or even cessation—of brain activity. In particular, there can be high mental activity without any corresponding high neural activity. What comes to mind are cases of Near-Death Experiences (NDEs) or Out-of-Body Experiences (OBEs) induced by G-LOC (or G-force in aerospace physiology, a loss of consciousness occurring from excessive and sustained g-forces draining blood away from the brain causing cerebral hypoxia), cortical deactivation through the use of high-power magnetic fields, mystical experiences induced through hyper-ventilation, and brain damage caused by surgery or strokes.

Many studies on NDEs suggest there can be mental activities without neural activities associated with flat-EEGs. Not surprisingly, all kinds of reductionist explanations have been proposed in reply: oxygen deprivation (anoxia), high carbon-monoxide levels, REM-sleep phenomena, psychedelic agents, hallucination. However, the question remains why not all people under circumstances of those types had ND-experiences. Besides, more research has been done in an attempt to rule these explanations out. A recent study by Sam Parnia suggests that NDE patients are "effectively dead," having no neural activities necessary for dreaming or hallucination; additionally, in order to rule out the possibility that near-death experiences resulted from lack of oxygen, Parnia rigorously monitored the concentrations thereof in the patients' blood, and found that none of those who underwent the experiences had low levels of oxygen. He was also able to rule out claims that unusual combinations of drugs were to blame because the resuscitation procedure was the same in every case, regardless of whether they had a near-death experience or not. If all of this remains standing, then mental states would indeed be different from neural states.

The power of mental states, independent of neural states, has also been backed by the study of *placebos*. A placebo is a "dummy" medication—an inert inactive substance, such as sugar, that is thought to have no effect on the condition being treated, and is given to patients to deceive them into thinking that they are receiving a substance that might affect an improvement or cure. Placebos are used in clinical medicine as well as laboratory research to test the effectiveness of new medications. Because there is a consensus among the medical and scientific establishment that "real" pain or "real" symptoms could only respond to an active chemical agent, the reasoning is that a chemical agent can only be proven to be effective if it "beats" an inactive agent in causing actual changes to pain or disease.

Nevertheless, it is evident that placebos—or actually a *belief* in their effects—are not really without effect. They do have an effect which is called the placebo effect. A placebo intervention may cause patients to believe the treatment will change their condition; and this very belief can produce a subjective perception of a therapeutic effect, causing the patients to feel their condition has improved. To put it briefly, just the belief that you are being treated can make you feel better, even so for a long time.

In a reductionist view, placebo effects are a nuisance, something that should not be. Sometimes doctors can explain these effects away as insignificant or a slight miscalculation in statistics. When this is not possible, scientists describe it as the placebo effect. The fact is, though, that the placebo effect stubbornly remains, no matter how small the number of patients, so that it is hard to deny placebos do have a "real effect." There is even a wide variety of things that exhibit a placebo effect: pharmacological substances, medical devices, sham surgery, sham electrodes implanted in the brain, sham acupuncture—all of these have exhibited placebo effects. Actually, the common practice of prescribing antibiotics for anything non-bacterial such as a viral cold or flu can be seen as an example of placebo use too. Perhaps the use of homeopathic medicines and hypnosis belongs to this category as well.

So far we discussed levels that most neuroscientists are willing to accept, although there may be lively discussion as to how they are related to each other. But the unanimity gets more fragile when it comes to the issue of whether there is one more level up—a hypothetical level *above* the cognitivity level of beliefs, reasons, and intentions. When the philosopher René Descartes described animals as pure machines, he declared a human being as a machine with a *mind*. Some neuroscientists—although a small minority—would argue that all beliefs, reasons, and intentions come ultimately from the mind, although their opponents would rephrase this concept as "a ghost in the machine" idea. It is time now to investigate whether the brain-mind distinction is relevant, or even acceptable, in this discussion.

A classical case for the existence of the mind—as an "agency" independent of the brain—was made by the neurosurgeon Wilder Penfield. Once Penfield asked his patient to try and resist the movement of his left arm that he was about to make move by stimulating the motor cortex in the right hemisphere of the brain. The patient grabbed the arm with the right hand, attempting to restrict the movement that was to be induced by stimulation of the right brain. Thus the left hemisphere was telling the right hand to restrict the movement of the left arm being forced to move by the stimulation of the right hemisphere. As Penfield said, "Behind the brain action of one hemisphere was the patient's mind. Behind the action of the other hemisphere was the electrode." The neurophysiologist Sir John Eccles concluded from experiments like Penfield's that "voluntary movements can be freely initiated independently of any determining influences within the neuronal machinery of the brain itself."

Those neuroscientists who posit the mind as an entity on its own level face some serious opposition. The problem begins with the term "mind." Opponents would counter that a thing such as "temperature" causes molecules to move about more rapidly, but only in a spurious way. If the term "temperature" offers an explanation with downwards causation that is only seemingly so, they say, because we have hypostatized ("thing-ified") an abstract term that does not really exist and cannot really do anything.

Is the mind perhaps also "something" that has been hypostasized as a powerless entity? Even if mental states and their contents such as beliefs, reasons, and intentions seem to be real actors in the causal story of human behavior, this does not necessarily entail that the mind is a real entity too, so the argument goes. Some have said rather compellingly that a mental event cannot cause brain processes to change behavior because it has no causal power of its own. What looks like downward causation is arguably just lower level causation of both the whole and the part. If my fear causes me to express more adrenaline, it is because both fear and adrenaline production are physiological—and hence physical—processes.

Nevertheless, according to some neuroscientists and many philosophers, there might be compelling arguments why the concept of mind cannot be "reduced away" by neuroscience. One argument goes along these lines. Those who identify the mind with the brain have to deal with the fact that thoughts and beliefs are not material entities like neurons. Thoughts are true or false, right or wrong, but never tall or short, heavy or light. If the mind were merely a brain issue, there would be no way to distinguish between true and false, or between right and wrong. Whereas the brain as a material entity has characteristics such as length, width, height, and weight, the mind cannot have any of those. If the mind were the same as the brain, thoughts could never be right or wrong and true or false, as neural events simply happen, and that is it! We can think about sizes and colors of things, but the thoughts themselves do not have sizes and colors. It should not surprise us then that people have known the contents of their own minds from time immemorial without knowing anything about brains.

Another reason why the distinction between mind and brain might be imperative, according to some, is the fact that it becomes hard, if not impossible, for reductionists to justify their own activities. If the mind were really a matter of neurons in the brain, it would belong to the material world and would therefore be as fragile as the material world itself. It would be sitting on a "swamp of molecules," unable to pull itself up by its bootstraps—which would be like an electric generator running on its own power. Nobel-laureate biologist Max Delbrück—who was trained as a physicist—described the contradiction in an amusing epigram when he said that the neuroscientist's effort to explain the brain as mere meat or matter "reminds me of nothing so much as Baron Munchausen's attempt to extract himself from a swamp by pulling on his own hair."

This argument seems to have quite some merit. In order to make any mental claims, even in science, we need to validate our claims as being true, otherwise they are worth nothing. If Watson and Crick, or Planck and Einstein, or any other scientists, were nothing but their neurons, then their scientific theories must be as fragile as their neurons. That would be detrimental to their claims. If we were nothing but a "pack of neurons," this very statement that we are making here would not be worth more than its molecular origin, and neither would we ourselves who are making such a statement. Concepts and thoughts behind a statement are mental, immaterial entities. To reduce them to a "creation of neurons in the brain" obscures the fact that "neuron" itself is an abstract concept. That would make for a vicious circle: The very idea that concepts are nothing but neurons firing is itself nothing but neurons firing. And the same holds for thoughts. To think differently is like saying that Shakespeare's thoughts are nothing but ink marks on paper.

The nearly legendary biologist J. B. S. Haldane reasoned along similar lines: If I believe that my beliefs are the mere product of neurons, then I have no reason to believe my belief is true—therefore, I have no reason to believe that my beliefs are the mere product of neurons. In other words, those who claim the mind to be identical to the brain are cutting off any reason they might have for relying on their own reasoning. It could be argued instead that when studying the human brain, neuroscientists make the human brain an object of science, which can only be done because they have a human mind that is the subject behind science. As Stephen Barr pithily puts it, "The brain does not infer the existence of the mind, the mind infers the existence of the brain."

Then there is a third argument to question the idea that mind and brain are identical to each other. Identity is more than correlation. A certain mental state may be correlated with a certain neuronal state, but that does not entail identity. The fact that two entities are correlated means that they must be distinct entities. It is true that my *mind* can be fully occupied with a specific problem. But it is either false or makes no sense to say that my *brain* is then fully occupied with that problem. If this is true, then my mind and my brain cannot be identical. The most we could say is that a certain part of the brain is thinking about a certain problem under consideration. Even if every mental event is a brain event, not every brain event is a mental event.

If there are brain states that are not mental states, then there must be some properties that distinguish these brain states that are mental states from the brain states that are not mental states. These properties will have to be specifically mental, as no physical property could do the trick. So it is very questionable whether our understanding of the world is done by the brain. Michael Augros uses the following analogy: You cannot count what you are seeing without using your eyes, but that does not mean your eyes are doing the counting. Similarly, it is clear that we cannot understand anything without using our brains, but it does not follow that our brains are doing the understanding. In order to observe, investigate, and understand the brain, there must be something "outside" the brain a neuroscientist studies. If there is knowledge there must be someone to know it. Our knowledge of the brain requires someone—a mind with an intellect—to know it. In other words, a known object always requires a knowing subject.

A fourth argument has been brought forward based on information theory and information technology. Crucial in information theory is the separation of content from the vehicle that transports it. In computers, as information theory shows, the content is manifestly independent of its material substrate. As George Gilder puts it, "No possible knowledge of the computer's materials can yield any information whatsoever about the actual content of its computations." What technologists call the "physical layer" is the lowest level of silicon chips and silica fiber on the bottom, whereas the programs and content are at the top.

Seen in this computer analogy, the brain with its neurons would only be the physical carrier of immaterial thoughts coming from the mind (see chapter 16). If the same thought can be transported by different vehicles—such as pen strokes on paper, currents in computers, impulses in the brain—the conclusion would be that the mental, a thought, is not identical to the neural, its carrier. Hence, the claim could be made that the brain does not create any thoughts but merely transports the thoughts coming from the mind. Needless to say that defects in the carrier, neurological or otherwise, may affect what comes from the mind—but would arguably not directly affect the mind. Each comparison falls short, but asking how the mental makes the neural work is like asking how a computer program causes the computer circuits to solve mathematical equations. A computer does not create its "own" activities but merely executes activities preprogrammed by its designer.

Reductionists would probably deem all the above arguments mere philosophical speculation. Interestingly enough, researchers at places such as the University of Virginia and the University of Vienna, Austria are studying a phenomenon that has been called "terminal lucidity"—the unexpected return of mental clarity and memory shortly before the death of patients suffering from severe mental disorders. This is the term used when dying people, who have previously been unresponsive or minimally responsive, suddenly gain clarity of mind for a few hours, often talking coherently with loved ones before passing away a short time later. We know that there is no observable sudden change in the brain when death is very near. Is it possible then that the mind's sudden and short-lived return to normalcy just before death is brought about, not by some inexplicable surge in brain functioning, but by the mind's distancing itself from the brain? Examples include case reports of patients suffering from tumors, strokes, Alzheimer disease, and schizophrenia. Although terminal lucidity has been reported for around 250 years, it has received little scientific attention because of its complexity and transience—or perhaps because it doesn't fit in a paradigm that equates the mind to the brain.

In response, one could possibly argue that the brain is as much responsible for thinking as the hand is for grasping or the leg for kicking. Many of the discussions about the brain's causality of thought seem to involve the idea that if one makes the brain responsible for thought, then it would somehow become the principle agent of thought. But this is as dubious as thinking that if one makes the hand responsible for grasping, that somehow it is the principle agent of grasping, as opposed to a mere tool used by a human being. What if it were the mind, instead, that uses the brain as its organ? This might suggest that the conscious self is an autonomous agent working through a material brain. The mind needs the brain to function properly, but the brain also needs the mind to function fully. The neural system may be a necessary, but arguably not a sufficient explanation of thinking.

Regardless of whether these arguments mentioned above are fully convincing, reductionists would still object to introducing something as immaterial as a mind. Doing so would be, in their view, a violation of ontological reductionism (see chapter 6). However, ontological reductionism is usually taken in its monistic form, which claims that all objects, properties, and events are reducible to a single entity—*matter*, that is—but theoretically there could also be a version based on *dualism* claiming that everything is reducible to one of two types of entities—either matter or spirit. Some argue there may be compelling reasons to go for the dualistic form of ontological reductionism. As said earlier, a one-level ontology contradicts itself, because the claim of a one-level ontology is in and of itself a matter of cognitivity. If so, we need at least a two-level ontology of cognitive events in addition to molecular events. (But there seems to be much more in between, which would call for a multi-level hierarchical structure.)

The idea of a two-level ontology was also accepted somehow by Norbert Wiener from MIT, for instance, when he defined, at the dawn of information theory in 1948, the new crisis of monistic reductionism in neuroscience: "The mechanical brain does not secrete thought 'as the liver does bile,' as the earlier materialists claimed, nor does it put it out in the form of energy as the muscle puts out its activity. Information is information, not matter or energy."

This and similar positions still leave the question open as to what the relationship would be between mind and body/brain. The mind is considered here as a non-physical entity that is nevertheless capable of exerting "downward causation" on the physical world, by putting constraints on the brain to further its goals. Many questions arise. How could a material brain ever carry out instructions given by an immaterial mind? Can the mind "direct" the brain like a programmer programs a computer? Is this not dualism in an unacceptable way? Indeed, many scientists would reject this kind, or any kind, of dualism. Their main reason probably is that dualism carries a heavy historical burden. It is especially in Cartesian dualism that mind and body are seen as completely different *entities*—immaterial and non-localized versus material and localized—and yet they are supposed to mysteriously interact—two different kinds of substances joined together. How would that be possible?

This Cartesian conception is connected with the so-called "homunculus fallacy," according to which consciousness is the work of the soul, or the mind—the inner entity that thinks and sees and feels, and which is "the real me inside." When Descartes compares our minds with a pilot in his ship, he seems to suggest that our minds are the pilot behind our eyes and behind everything our body does. Critics reject this idea because it casts no light on the consciousness of a human being by simply re-describing it as the consciousness of some inner little being [*homunculus*]. So they replaced this mysterious being with the monism of the brain.

But as Max Bennett and Peter Hacker have argued, this homunculus fallacy keeps coming back in another form. Now the homunculus is no longer a soul, but a brain, which is supposed to "process information," to "map the world," to "construct a picture" of reality, and so on. Oddly enough, this brings dualism back in a hidden way, for all these latter expressions can only be understood because they describe conscious processes with which we are already familiar—processing, mapping, and constructing. To describe the resulting form of "neuroscience" as an explanation of consciousness is dubious, as it merely reads back into the explanation the feature that needs to be explained.

This might bring the mind-body dualism back into the discussion. Some dualists nowadays take body and mind as two different *aspects*—the mental and the physical—or two different *dimensions*—matter and form—of the same human being; one can *tell* them apart but one cannot *set* them apart. The brain belongs to the world of objects, whereas the mind is part of the world of subjects. In this view, body and mind are not two individual substances, like in Cartesian dualism; rather are they a unity whose nature is comprised of both material and immaterial form, so the body becomes what it is due to the mind.

The fact that we distinguish body and mind does not entail that we can separate them, any more than the idea of a three-dimensional space means that we can separate those three dimensions. Someone like Stephen Barr compares this with the relationship between mass and gravity; we do not really know how these two interact. Or take the case of electrically charged particles that interact with each other through the mediation of electromagnetic fields; the charged particles affect the fields and the fields affect the particles, but we do not know anything about the "mechanism" behind this interaction. Something similar might be said when it comes to body and mind: We know *that* they interact but do not know *how*.

For those who do accept the notion of mind as a higher-level entity, it would be possible to argue that mental events have the "power" to change behavior, even though they are not merely neural events but come from the mind instead and belong to the realm of cognitivity, thus making for their own level of existence. If this claim is valid, we would have to sharpen our terminology: Neuro-scientists are not mind-readers, neuro-surgeons are not mind-surgeons, and neuro-science is not mind-science. But that might be hard to accept for many neuroscientists of the reductionist camp.

Yet, although it may not be mainstream thinking, some very sophisticated scientific and philosophical thinkers—such as the Nobel laureate and neurobiologist Sir John Eccles, the philosopher of science and religion Richard Swinburne, and the philosopher of science Sir Karl Popper—are mind-body dualists who take the reality of the mental most seriously. But of course, this discussion cannot be decided based on "authorities."

For further reading:

Churchland, Patricia Smith. *Brain-Wise: Studies in Neurophilosophy*. Cambridge, MA: MIT Press, 2002.

Cosmides, L., and J. Tooby. "From function to structure: The role of evolutionary biology and computational theories in cognitive neuroscience." Pp. 1199-1210 in *The Cognitive Neurosciences*, ed. M. S. Gazzaniga. Cambridge, MA: MIT Press, 1995.

Gazzaniga, Michael S. *Human: The Science Behind What Makes Us Unique*. New York: Harper Perennial, 2009.

Sperry, Roger W. "Science and moral priority: merging mind, brain and human values." Vol. 4 of *Convergence*, (Series ed. Ruth Anshen) New York: Columbia University Press, 1982.

14.   Sociobiology

Social scientists who use a reductionist approach to explain animal and human behavior are in search of explanations located at lower hierarchical levels. It is not very likely that they come up with explanations in terms of quarks and the like, but they do go as far down as to the level of genes. This approach has become known as sociobiology—the invasion of evolutionary biology into the domain of social and behavioral sciences. What G. C. Simpson did for paleontology, sociobiologists such as E. O. Wilson try to do for social behavior.

Sociobiology is the systematic study of the biological, and more specifically genetic, basis of social behavior, including its evolution. It deals with interactive social traits—not only in the animal world but also in human societies—traits such as cooperation, aggression, offspring care, altruism, territoriality, warfare, slavery, kinship rules—the list goes on and on. Sociobiologists try to explain all of these in a genetic and evolutionary context through a cost-benefit analysis (see chapter 11). The sociobiologist E. O. Wilson, for instance, insists that human rules against incest have a genetic basis.

Some serious objections have been made against such an approach. First of all, certain sociobiological claims come close to a circular argument: The cultural universality of incest suggests a genetic basis; since there is a genetic basis, incest is a universal phenomenon in human cultures. Second, it is common practice among evolutionists to divide the similarities between two species into *homologous* features shared by common descent and a common genetic constitution, versus *analogous* traits which evolved separately. Claims about a genetic similarity can only be made if they are based on homologous traits. But how can we know if similarities are homologous or analogous, for analogous features can be strikingly similar? So it is hard to tell whether a certain feature such as incest is genetic or merely an analogous pattern that might be purely cultural in origin. If it is cultural, natural selection has no genes to operate on.

To analyze further criticisms, we will focus mostly on the sociobiological explanation of what is called altruistic behavior in animals and humans, since this seems to pose an acute problem in terms of natural selection. Let us call this the problem of *bio*-altruism. Why is this form of altruism a potential problem for socio-biologists? Well, altruism seems to defy the very idea of natural selection. Altruism is considered to be unselfish behavior—a sacrifice of personal comfort for the *benefit* of others—whereas natural selection is based on the principle of increasing one's own reproductive success at the *expense* of others. So the pressing question for socio-biologists is the following: How can altruistic behavior in the animal world still be advantageous to its agent? In response, socio-biologists such as E. O. Wilson and W. D. Hamilton claim to have found a biological and evolutionary explanation. Let's see how.

In the animal world, we do indeed find many examples of animals helping one another; just think of sterile worker bees "unselfishly" helping the queen raise her own progeny. Sociobiologists would explain bio-altruism as a form of helping one's close relatives, because these carry DNA very similar to one's own. This could be called "enlightened" self-interest—in helping others, one helps oneself. Since natural selection is a matter of balancing the benefits against the costs, bio-altruism is a way of promoting one's "own" DNA by diminishing one's own offspring (a cost) but increasing the offspring of one's relatives (a benefit) who have similar DNA. And that is exactly what bees accomplish. Because these *Hymenoptera* have a very peculiar sexual system, females are more closely related to sisters (sharing 75% of their genetic material) than to daughters (sharing only 50%). So sterile females actually increase their own reproductive success by 25% in an "indirect" way—by helping rear their queen-sister's progeny (75%) rather than their own (50%).

This biological approach has even wider implications. As the geneticist J. B. S. Haldane once said, "I will lay down my life for two brothers or eight cousins." His calculation of costs and benefits was based on the fact that in a sexual system such as ours, siblings share ½ of their genetic material, whereas cousins only share ⅛. By helping close relatives, one is somehow promoting dispersal of one's own genetic material, but in an indirect way through relatives—for instance, through two brothers (2 x ½=1), or 8 cousins (8 x ⅛=1). This might be called altruism, but it is safer to label it as *bio*-altruism.

The more technical terms sociobiologists use to explain phenomena like these are *kin selection* and *inclusive fitness*. Kin selection is the evolutionary strategy that favors the reproductive success of an organism's relatives, even at a cost to the organism's own survival and reproduction. Kin selection is an instance of inclusive fitness, which combines the number of offspring with the number an individual can produce by supporting others, particularly siblings. Such an individual would reproduce vicariously rather than directly, so to speak. According to Hamilton's rule, kin selection causes genes to increase in frequency when the genetic relatedness of a recipient to an actor multiplied by the benefit to the recipient is greater than the reproductive cost to the actor.

Interestingly enough, sociobiology seems rather reductionist in its approach when seen from a sociological perspective, but seen from an organism's level, it could be labeled as holistic since it considers the individual as part of a wider whole. In the concept of kin selection, for instance, *kin* is not the name of a super-entity which *replaces* individual organisms in the selection process, but a pointer to the necessarily *social* character of some behavior. This social character can cover various ranges: Parental care helps mainly one's kin; the mobbing of predators helps primarily one's group; migration may help mostly one's species. In other words, inclusive fitness theory pertains principally to genes underlying behaviors that influence other individuals. Although it may be handy to assume that individual organisms generally act in ways that maximize their inclusive fitness, inclusive fitness is not a property of individuals—it represents the cumulative effects of genes that underlie specific behaviors in a wider social setting.

In addition to bio-altruism, there is also a sociological version of altruism—which we should call *socio*-altruism, in order to keep our terminology clean. It is based on the principle of helping those who return the help; divided you may fall but united you may conquer—something like "I give so you give." This happens, for example, when two or more organisms, such as chimpanzees, band together, and thus in helping others help themselves. Since there is social behavior in the animal world, there may also be socio-altruism. Sociobiologists have studied this phenomenon with what they call "game theory," and they have been rather successful doing so. The "Hawk-Dove" game, for instance, refers to a situation in which there is a competition for a shared resource, and the contestants can choose either appeasement or conflict. The principle of such games is that while each player prefers not to yield to the other, the worst possible outcome occurs when both players refuse to yield.

Where does this leave us? The reductionist approach of sociobiology may have been successful to a certain extent, but from a holistic perspective, there are some questions left, especially when it comes to human societies. First of all, the inclusive fitness theory does not imply that all genes under all circumstances are driven to maximize copies of themselves at the expense of non-copies. Holists would point out that there are emergent properties in human behavior that cannot be reduced to the maximization of inclusive fitness. Weaning, for instance, makes a case in point.

Weaning is the process of gradually introducing an infant to what will be its adult diet and withdrawing the supply of the mother's milk. Weaning has several levels of explanation: The first level is the one of sociobiology, the second one of physiology, the third one of individual experience and psychology, the fourth level is one of socio-cultural organization. A one-level analysis may not be very adequate in cases like these. There are variations in weaning behavior that cannot be explained by the lower levels of explanation. The main issue about the nature of causation in these fields has to do with the question of whether explanations of social phenomena are fully reducible to explanations of the individuals involved in social phenomena (more on this in chapter 15).

A second issue, seen from a holistic perspective, is whether bio-altruism and socio-altruism do actually cover all forms of altruism in human society? The answer seems to be negative. In addition to bio-altruism and socio-altruism, there is also *moral*-altruism, which is behavior for the sake of the moral *value* of serving others, without expecting any direct or indirect advantage. This is what most people consider altruism to be— "unselfish altruism." Moral-altruism may very well be partly based on the biological features of socio-altruism and cooperation, but it also goes sometimes far beyond it. Moral values, moral laws, or a moral code—or more in general, morality—do not occur at a biological level—as little as cognitivity does. The notion of charity, for instance, is completely about giving for giving's sake. Or take donating blood to strangers: It does not help relatives, nor does it help banding gang members; it is neither bio-altruism nor socio-altruism; plus it is not subject to natural selection. When firefighters or soldiers on the battle field die in the line of duty, they typically do not give their lives for their relatives' sake.

Put differently, moral-altruism does not happen "by nature." If morality were anchored in the genes, why would we need an articulated moral rule or code to reinforce what "by nature" we would or would not desire to do anyway? That does not make any sense. If morality is encoded in the genes, a moral code would be completely redundant. If moral behavior were genetic, there would be no need for a moral code as well. Bees have the genes for it, but no moral code; we, on the other hand, do have a moral code but without any corresponding genes.

Apparently, moral-altruism is a form of altruism that is guided by a moral code at a higher organizational level. It is actually in constant battle with our urge to survive, to care for ourselves, and to provide for our offspring. In short, moral-altruism is no good friend to survival; it amounts very often to "genetic suicide." In general, moral laws do not have any survival value and therefore cannot be promoted by natural selection. Morality and "survival of the fittest" do not go well together. Natural selection is about success at the *expense* of others; morality is about duties to the *benefit* of others. Francis Collins, the former Head of the *Human Genome Project* and currently Director of the *National Institutes of Health*, put it very bluntly, "Evolution would tell me exactly the opposite: preserve your DNA. Who cares about the guy who's drowning?"

In other words, moral-altruism is not a biological but a moral concept that is related to moral values. It should be clearly distinguished from the other two forms of altruism for the following reasons:

1. Bio-altruism is behavior with the *effect* that one's own offspring is diminished but compensated for by helping relatives. So it may be functional and advantageous at times.

2. Socio-altruism is behavior with the *motive* of helping others, but limited to those who return the help. In helping others, one helps oneself according to the Roman motto *Do ut Des*.

3. Moral-altruism is behavior for the sake of the moral *value* of serving others, without expecting any advantage. This is what we really and rightfully call "unselfish altruism."

So it is important to keep our terminology straight. What one *ought* to achieve (which is a moral value or rule) is not necessarily what one *wants* to achieve (which is a motive or intention). And what one wants to achieve is not always what one actually *does* achieve (which is a mere effect). The following simple question might further clarify the difference between a biological, social, and moral approach: What is wrong with rape (or any other kind of violent behavior)? As a biological feature, it may be very advantageous to have such a gene or allele. As a social strategy, it may be a very effective strategy of banding together against someone else. But as a moral issue, it is generally considered plainly wrong.

When sociobiologists such as E. O. Wilson define an act of altruism as one that occurs "when a person or animal increases the fitness of another at the expense of his own fitness," they are specifically ignoring the distinction between acts that merely have the *effect* of benefiting others and acts that are performed with the *intention* of, let alone the *value* of, benefiting others. Similarity of result does not imply identity of cause. By ignoring this distinction, they are likely to miss the entire point of labeling a piece of social behavior or moral behavior as altruistic. Without acknowledging intentions, we could not even tell between a suicidal death and an accidental death. Of course, a radical reductionist could always make the objection that people may *think* that their actions are based on values and intentions, but that they in fact are determined by genes. However, such an objection works like a boomerang; it would also make itself the mere outcome of genes. Would those who disagree on this subject merely have different genes? It's hard to believe.

A third limitation of the reductionist approach in sociobiology is that even animal behavior is not completely genetically programmed. This holds even more so for human social behavior, which does not consist of rigidly determined, fixed patterns that are uniform throughout the species; it rather varies from one social group to another—not to mention from one individual to another. Human behavior can show actions that are different from what is genetically programmed. Take just a simple case like left- or right-handedness. It most likely has a genetic basis, yet it can be changed by what is culturally preferred and socially sanctioned. Most genetic behavior programs are "open" to additional input from non-genetic sources—that is, they can be "socially learned."

In other words, behavioral *genotypes* never account for all the variations in behavioral *phenotypes*—especially not in human beings. Even in very simple organisms, adult behavior repertoires vary according to each organism's learning history. It must be granted that some sociobiologists do agree that most human social responses are socially learned, but they still believe that cultural variations are merely a consequence of a genetically programmed *scale* of alternatives that can be turned on and off by environmental switches—a "behavioral scale" in Wilsonian terms. The scaling would explain why some human populations are cannibals while others are vegetarians, why some are polygynous while others are polyandrous, and so forth. This has been seen by some, though, as a desperate move to hold on to a reductionist approach.

A fourth holistic issue would be the impact of cultural factors and norms on human behavior. There are definitely cross-cultural variations of behavior. A theory of human evolution must also take into account the individuals' incessant influence on themselves in the form of cultural creations. Humans create themselves in whatever image they have of themselves. The social sciences are typically interested in explaining these variations (see chapter 15). Subsuming all of them under a common biological or genetic header will not explain the widely disparate range of behaviors that are *sociocultural* in nature. Major cultural changes have occurred in a period of time far too rapid to have been caused by genetic changes and natural selection.

The problem with such biological reduction is well worded by Michael A. Simon, "In order for a human trait to be explained biologically, it must first be 'biologized' [...] is that it is likely to sacrifice precisely those features of human social behavior that give it a socially or philosophically distinctive character." So by "biologizing" morality, we inevitably lose its distinctive moral character. Let's use the following example to show what we could lose in the process.

Even if it is custom in certain societies that the adult male responsible for care of children of any particular woman is not the husband but the woman's brother(s), this may be explained by the fact that an uncle is more closely related to the child than the father whose paternity is in doubt, thus making undoubted relatives more preferable, in a sociobiological sense, than those who may not be related to the child at all. However, this reasoning does not seem to explain why not all, or at least most, societies have childcare done by uncles instead of fathers, as cultural anthropology shows us. Somehow, sociocultural factors at a higher level come into play as well. Even if the custom of uncles taking care of children occurs more often in societies with much extra-marital sexual intercourse, the question remains where the preference for this kind of intercourse comes from. It seems to be hard to consider only genetic factors.

Something similar could be said about incest taboos. Even if it is their biological function to prevent potentially deleterious inbreeding, this would not exclude the possibility that they serve sociocultural functions as well. In fact, the "rules" as to what is considered incest can vary considerably between cultures. The Pharaohs in Egypt, for instance, had their own specific "rule." The inadequacy of sociobiology's way of explaining evolution can be seen in its attempt to reduce cultural evolution to mere biological evolution. Natural selection may be powerful, but probably not all-powerful. As some have said, explaining the origins of culture by reference to evolutionary concepts is like explaining the origin of the steam locomotive by referring to the application of gas laws.

Nevertheless, sociobiology has had quite an impact on the reductionists' agenda, even in psychology. However, if sociobiology really has the last word, we end up with all sorts of survival-inducing illusions: social illusions, political illusions, moral illusions, religious illusions, etc. No wonder some have grown to be highly critical of what has become known as evolutionary sociology or evolutionary psychology, because their advocates seem unwilling to draw lines between what can be taken as demonstrated and what remains speculative, making these disciplines a matter more of faith than of science. As the biologist Jerry Coyne puts it, "If we are truly to understand human nature, and use this knowledge constructively, we must distinguish the probably true from the possibly true."

For further reading:

Alonso, W. "The role of Kin Selection theory on the explanation of biological altruism: A critical Review." *Journal of Comparative Biology*, 1998, 3 (1): 1–14.

Montague, A. ed. *Sociobiology Examined.* Oxford: Oxford University Press, 1980.

Rose, Steven, Leon J. Kamin, and R. C. Lewontin. *Not in Our Genes: Biology, Ideology and Human Nature.* New York: Pantheon Books, 1985.

Wilson, E. O. *Sociobiology: The New Synthesis.* Cambridge MA: Harvard University Press, 1975/2000.

15.   Social Sciences

The term "social sciences" is used in this chapter as a collective term for the various branches of science that attempt to understand human society and human behavior—ranging from sociology, anthropology, and psychology to economics and political science. It was Émile Durkheim who developed the notion of "social facts" to delineate a unique empirical object for the social sciences to study.

As we have seen, reductionism can work at different levels. When applied to the social sciences, the lowest level of reduction offers physiological explanations which attempt to explain human behavior in terms of chemistry, genetics, and brain structure. At the highest sociocultural level, explanations focus on the behavioral impact of where and how human beings live. Between these extremes there are behavioral, cognitive and social explanations.

We encountered these levels already in our discussion of neuroscience (see chapter 13). Traditional neuroscience has for many years considered the nervous system as an isolated entity, thus largely ignoring influences of the *sociocultural* environments in which humans live. Currently, neuroscientists are beginning to recognize the considerable impact of social and cultural structures on the operations of brain and body. Not only our language, but also our voices, gestures, habits, and beliefs do not just come from ourselves but also from the society and culture we are surrounded by.

In this chapter, we will focus on the *social* and *sociocultural* dimension of all human behavior—a dimension from which physicists and biologists typically "per definition" abstract. Since humans are fundamentally a social species, rather than individualists, they seem to create emergent organizations beyond the individual—structures that range from nuclear families, extended families, and clans to cities, tribes, civilizations, nations, cultures, and societies. The question is, of course, how "real" these high-level structures are.

Let us begin with the society "as a whole." The society seems to have decisive effects on the individuals. John Stuart Mill once demonstrated that it is never open to us to behave as we please, not even in the most liberal society. However, society does more than restrain individual actions. It also impacts our thinking and feeling. Each individual life, for instance, is excessively conditioned by language. Well, language is thoroughly dependent on society. Additionally, the knowledge each individual has is acquired mostly by tradition and from hear-say—that is to say, it has come through society. Even personal emotions and desires are often a result of upbringing—that is, of the feelings and opinions society at large entertains. No wonder that society has always appeared as a power of intense actuality. Society seems to be something existent, something real, just like all other real things in the world.

Immediately, however, problems pop up. Looking around in society, we perceive only human beings—individual human beings, that is. Terms such as "society" and "humanity" seem to be just simple collective terms for all, or at least certain, people combined; one never encounters a society as such (we saw a similar problem with a biological species in chapter 11). Although society faces us as a real and actual power, it cannot be situated anywhere; there doesn't seem to be such a thing in the world. These considerations have persuaded some philosophers and sociologists to voice the view that "society" is a pure fiction. They take it that only individual persons exist in reality—a collection of autonomous, pure egos without any ties or connections.

Some call this the Crusoe model of an "isolated man face-to-face with nature." In this model, society only exists by the grace of individuals in the same way an ant colony only exists by the grace of individual ants; without them, the society or colony would cease to exist. When talking about a nation, for instance, one does not really refer to a nation but to its citizens, or more often, to those among them who are in charge. Consequently, something like duties regarding the nation are actually duties regarding specific individuals—the head of the nation, federal agents, civic leaders, and so on.

However, this view raises questions of its own according to its critics. How can proponents of this kind of individualism—let us call them "individualists"—ever explain the evident fact that there is some kind of pressure that society seems to exert on each one of its members in the form of legal laws and the like?

It must be said, though, that individualists don't really deny this evident fact. They are even able to explain it by declaring that this pressure is a consequence of the mutual interactions individuals engage in with one another—which opens the door for a very basic kind of holism.

Nevertheless, the crucial point is that in the individualists' view, the individuals are seen as the only real entities in society, making up the whole all by themselves. Society is nothing but the sum total of all individuals. In other words, interactions between people are in essence "unreal." Were these individualists to take interactions as actual and real, then they could never claim that all there is to society is mere individuals. If individualism, as we described it here, were true, it would be impossible for society to claim any duties from individuals, because interactions between individuals are seen as merely fictitious.

Society, in the individualists' view, does not have "interests" and does not aim at anything, so it is incorrect to say that individuals are working for the common good, because collectives do not have ends or desires. This view of society "atomizes" society, so that the single individuals act like atoms and molecules in gases do—like "atoms" next to one another or bumping into one another. It leads to Thomas Hobbes' "war of all against all" and causes a man to be "a wolf to his fellow man."

The other extreme viewpoint in this discussion is that society is the one and only real entity, at the "cost" of individuals—existing as some kind of "super-organism." Consequently, the individuals are seen only as parts of a larger whole, and therefore do not qualify as complete entities. Just like the hand of a person is not complete in itself, but only a part of the person, in the same way an individual person is considered only a part of society. As a consequence, individuals have no rights and duties of their own. After all, they live their lives as part of society, through the power of society, and for the benefit of society. The result is some kind of socio-ethical collectivism, ranging from authoritarianism (in which a single power monopolizes political power) to totalitarianism (in which the regime attempts to control virtually all aspects of social life). What they basically have in common is that they turn citizens into a mere means for the sole purpose of advancing society—in the same way ants and bees are believed to exist only to keep the colony alive.

Somehow the aforementioned one-sided views on society are the outcome of either an extreme reductionist or holistic presupposition. They represent a specific position in the ontological reductionism debate. They accept only the individual or only the society or only the interactions between individuals as the basic entity for the social sciences (see chapter 3). But usually, social scientists take a position somewhere between these extremes. They are more interested in methodological reduction, pro and con, than in its ontological version.

Let us see how this rather abstract discussion pertains to the social sciences. A methodological individualist is a reductionist who maintains that the right way to approach the study of a society is to focus on the behavior of the individuals that compose it. A methodological holist, on the other hand, believes that such an investigation will fail to shed much light on the nature and development of society as a whole with its rules as to how individuals interact with each other. Both approaches are arguably true, because social phenomena are the result of organisms evolving at both the individual and social levels of biological organization at once. And both would be false, if each takes the truth of what it is defending to deny the truth of what the other side is defending. A middle road would be to focus on interactions between individuals within the setting of a society.

The viewpoint of pure individualism has in fact only very few proponents. Thomas Hobbes, Adam Smith, Max Weber, and more currently F. A. von Hayek, J. W. N. Watkins, and Karl Popper are individualists, at least in a methodological sense, because they think of individualism as the key principle about how to practice social science. Ayn Rand, who promoted the "virtue of selfishness" as well as a form of laissez-faire capitalism, could probably be considered an individualist in this sense too. Hayek worded his individualism this way, "the individual's system of ends should be supreme and not subject to any dictation by others." Another representative, Ludwig von Mises of the Austrian School of economics, would also argue that only individuals act, and that society does not exist apart from the thoughts and actions of individuals, as society is nothing more than people cooperating, each to achieve their own individual interests. Hence, society is not a *sui generis* entity, let alone a "super-organism." As Mises writes, "there is I and you and Bill and Joe and all the rest."

More recently, sociobiologists have joined this group, because it is their aim to explain social phenomena by natural selection working on the individual level (see chapter 14 as to whether this is entirely true). What all individualists have in common is their conviction that all social phenomena can be explained, in principle, as a consequence of the behavior of the individuals involved in the situations they face. It does not necessarily entail reductionism to levels farther down the hierarchy, as there is no need or desire to further reduce social activities to quantum events. Whenever problems in the human sciences remain unsolved, they argue, it is because it is hard to elucidate their complexities—certainly not because we don't understand subatomic physics well enough. It was already one of the founders of sociology, Max Weber, who defended the position that sociology is a science, but not a natural science.

In contrast, a more holistic oriented approach holds that theories in the social sciences must include unobservable theoretical entities, such as social structure. From an ontological viewpoint, it is believed that such entities "really" exist in some way that is not reducible to the individuals and their behavior. This is the core of a "system" approach. As the psychologist George A. Miller puts it: "To a reductionist, a systems account is not an explanation; it is something to be explained. But reductionistic accounts of human institutions or human language seem doomed to failure. [...] Analyzing physical events to find explanations for social conventions is like analyzing air molecules to explain the wind." In a simple case like a traffic jam, for instance, system analysts would counter that this cannot be explained through any given drivers trying to get to work; it is rather the overall laws of the "collective" that explain the situation—in spite of, not because of, the behavior of each individual. Their main point is that any study of social phenomena is more than a mere summation of the facts of individual behavior.

A more moderate version of holism would focus on the interactions between individuals. Although it does not assume a *"super*-individual" above and beyond the individual members of a society, the society itself might still exhibit *supra*-individual properties and entities, not to be found in its members. Structural functionalism, for instance, is a sociological paradigm which addresses which social functions various elements of the social system perform in regard to the entire system. In addition, there is a sociological approach called interactionism. It is a theoretical perspective that derives social processes—processes such as conflict, cooperation, and identity formation—from human interaction by studying how individuals interact within a society.

It is easy to imagine why sociologists are often holistic thinkers. It is very tempting to see society as some kind of organism. In a real organism, there is some form of "communication" between its parts through the blood stream and the nervous system. In a similar way, there is "communication" between the members of a society through language, books, schools, mass media, social media, and the internet. Many social scientists focus on such interactions between individuals. That's why the social sciences are often defined as an academic discipline concerned with society and the relationships among individuals within a society. Whenever there is a multitude of individuals interacting with one another, there often comes a moment when disorder gives way to order and something new emerges: a pattern, a collective decision, a common structure, or a change in direction—albeit ultimately rooted in activities of individuals.

This kind of phenomenon can even be observed among animals. Flocking behavior, for example, is the behavior exhibited when a group of birds, called a flock, is foraging or in flight. There are clear parallels with the shoaling behavior of fish, the swarming behavior of insects, and herd behavior of land animals. It is considered the emergence of collective behavior arising from simple rules that are followed by individuals and does not involve any central coordination. Scientists have demonstrated a similar behavior in humans. In their studies, people exhibited the behavioral pattern of a "flock": If 5% of the flock would change direction, the others would follow suit. When one person was designated as a predator and everyone else was supposed to avoid him or her, the human flock behaved very much like a school of fish.

As a matter of fact, groups of human beings, left free to each regulate themselves, tend to produce spontaneous order, rather than the meaningless chaos often expected. A classic traffic rotary is an example of this phenomenon, with cars moving in and out with such effective organization that some modern cities have begun replacing stoplights at problem intersections with traffic rotaries, and getting better results. Emergent social processes like these can be seen in many other places and situations, such as cities and economic market places. They can be looked at as self-organizing phenomena. Translated in political terms, self-determination of individuals is better for society than regulation by the government—which has become known since 1776 as the "American Experiment." Phenomena like these occupy an odd position in the reductionism discussion because components are used to explain the behavior of the system, but it is the nature of their interactions (not their specific characteristics) that generate patterns of behavior, which are often referred to as emergent properties of the system.

An interesting application of this idea is the game theory as developed by the Nobel Laureate, mathematician, and economist John Nash, who became widely known through the Movie "A Beautiful Mind." The simple insight underlying John Nash's idea is that one cannot predict the result of the choices of multiple decision makers if one analyzes those decisions in isolation. Instead, one must ask what each player would do, taking into account the decision-making of the others. So what looks like a game done by mere individual players turns out to have a holistic dimension.

The stock market (or any market for that matter) is another example of emergence on a grand scale. As a whole it precisely regulates the relative security prices of companies across the world, yet it has no leader. When no central planning is in place, there is no one entity which controls the workings of the entire market. Agents, or investors, have knowledge of only a limited number of companies within their portfolio, and must follow the regulatory rules of the market and analyze the transactions individually or in large groupings. Yet, trends and patterns emerge. This phenomenon was coined by Adam Smith as the "invisible hand." While individuals act in their own self-interest, the result is that society benefits, and no government action is needed to insure that the public interest is promoted. It is private greed that leads to public benefit, says Smith.

Methodological individualists point to the market as their prime example of the "invisible hand" at work—of how the rationally explicable behavior of many individuals in the situations they face has consequences that none of them may intend. This may seem a clear case of reductionism, but it has some holistic elements that individualists easily overlook. When Adam Smith explains the market in terms of an "invisible hand," he implicitly assumes that the market is a system based on rules of interaction. If we lived in a world where the macro level would simply reflect the micro level, says Andy Denis, "There would be nothing for an invisible hand to do. The individual would be directly social, or, what comes to the same thing, there would be no separate category of the social. [...] A reductionist world would be a laissez-faire world." And then he concludes, "The invisible hand is what allows us to think, and act, in a reductionist way in a holistic world: it underpins reductionism by tacitly conceding holism. [... It] allows us to assert that, in practice, the world can be treated *as if* it were reductionist." [Italics added.] In contrast, an economist such a John Maynard Keynes would clearly acknowledge that economics should be "chiefly concerned with the behaviour of the economic system as a whole."

Not only does a reductionist approach of social phenomena focus on the level of individuals, but often even farther down to the level of the genes of individuals. It often tends to explain complex social phenomena, such as violence, alcoholism, or the gender division of labor in terms of disordered molecular biology or genetics. The reasoning is simple: If there are criminal acts, there must be criminal genes; if there are altruistic acts, there must be altruistic genes; and so on. But the critics of this approach would point out that the laws of genetics are not the laws of learning or the laws of speaking or the laws of social living. These laws occur at different levels, to begin with. Although human behavior does have a genetic basis, other factors may interfere at a higher level of organization, ranging from very simple learned behavior to rather complex creative behavior and social behavior. Therefore, a holistic approach of the social sciences would also look for factors of downward causation. What social scientists often call "the autonomy of the social" is the impact of social causes operating at a level different from the psychological and biological level and deriving from genuinely social forces.

This line of thinking is still consistent with what one of the founders of sociology, Max Weber, had expressed much earlier when he described sociology as "the science whose object is to interpret the meaning of social action and thereby give a causal explanation of the way in which the action proceeds and the effects which it produces. By 'action' in this definition is meant the human behavior when and to the extent that the agent or agents see it as subjectively meaningful."

Let us return to the supra-individual properties and entities in society that holism would speak of. When talking about these properties and entities, another often mentioned concept is culture—especially in a field like cultural anthropology. That is why supra-individual properties and entities are often combined in the adjective sociocultural. Individualists tend to assume that contributions to culture come from individual geniuses who bestow their insights on the rest of us. That is, the culture is supposed to evolve by the rational selection of arguments by individual rational subjects in society. Yet, this view obscures the fact that their insights are in turn partly the result of the culture they live in.

Most people are not aware of the impact culture has on them until they move to another country with a different culture. Culture is a supra-individual system that individuals are surrounded by like fish in water. It is basically "gene-free"; its features can be acquired or wiped out within the life span of a single generation, without any reproductive episodes taking place—so it is hard to conceive of it as just a biological system. Marvin Harris rightly remarks, "Edison's invention of the phonograph would have spread around the world even if Edison and all his close relatives had been entirely childless."

As a consequence, human populations and their members can in fact acquire every conceivable aspect of the world's cultural inventory. Sociology has nothing to fear from sociobiology if there are no related genes involved for natural selection to work on. But even if there are genes involved, they do not turn social arrangements into something biologically inevitable. Our genetic makeup permits a wide range of behaviors; it is due to sociocultural factors that we place a restriction on the wide range that our genes permit. We are not at the mercy of our genes and our genes are not our destiny; as it is often said, our genes are like a hand of cards we are dealt, but how we play them is up to us.

One way in which the level of society affects the individuals as rational beings is through cultural evolution—clearly to be distinguished from biological evolution. It is a "method" of transmitting features to later generations that does not involve genes or sex cells or kinship. Each individual internalizes the culture of society as a normal part of development after birth, including not only the language and the capacity to generate arguments, but also the arguments and conclusions that have accumulated in the history of society. Even the size of social groups—nuclear families, extended families, clans, tribes, peoples, races, cities, states, nations, social clubs, and so on—as well as their characteristics are not only defined internally but also by downward causation from a higher level, the level of culture. It is culture that partly delineates who belongs to these particular social groups, and who does not. Also how people react in polls and on the stock market depends on the impact of the specific society and culture they belong to.

This is the reason why correlations and associations found in social surveys and the like often surpass a one-level analysis. The fact, for instance, that statistics have shown that hurricanes with female names cause more casualties than those with male names cannot be explained only on one level, because there is no direct connection between the two. The explanation has to make a "detour" through a higher level of cognitive beliefs that make people expect—in part, culturally determined—that hurricanes with female names are less violent, so fewer precautions are being taken. Something similar holds for the particular units of measurement a society uses—meters versus inches, pounds versus kilograms, liters versus gallons. As the cyberneticist Francis Heylighen puts it,

"The choice of units is not determined by the laws of physics, but is the result of a complex socio-cultural evolution in which different units are proposed for the most diverse reasons, after which one unit is eventually selected."

In this context, some have used the concept of *meme*, inspired by developments in sociobiology (see chapter 14). Coined by Richard Dawkins, it stands for a supra-individual component of culture. Just as genes are the basic unit of biological fitness, so memes are seen as the basic unit of cultural fitness. Some memes, such as tunics or sorcery, have their day and then die out. Others, such as science and democracy, last for centuries. Whereas a *gene* for celibacy, for instance, may be doomed to failure in the gene pool, a *meme* for celibacy can be quite successful in the meme pool.

Memes, like genes, compete for the long haul. As Daniel C. Dennett puts it, "Whatever you contribute to your culture may still live on long after your genes have dissolved in the common gene pool." In this view, cultural selection favors memes that exploit their cultural environment to their own "advantage." Few remember popular books from decades ago, but the ideas of Aristotle are still with us. Aristotle may or may not have many of his genes alive in the world today, but the meme of Aristotle is still going strong.

The rise of civilization, in this light, can be seen as the ever increasing efficiency in meme replication. In ancient times, the only way a meme could be passed on was through word of mouth, then came writing, and now the internet. But what has remained the same, in this view, is that memes drive societies to improve themselves. They somehow control the members of a society in a holistic way. Powerful ideas do matter—they tend to transcend the coincidence of time and place. It is here that the level of culture and the level of biology clearly separate. Take the Ebola crisis as an example. The fear for Ebola spreads much faster through the media than the Ebola virus itself ever could through physical contact. Apparently, in addition to natural selection, there is something like cultural selection. Similar to Darwin's natural selection theory, cultural selection theory has three phases too: variation, reproduction, and selection. Variation gives rise to a meme, reproduction is responsible for its spread, and selection is based on the factors that control the spread.

Another important part of culture is *language*. Spoken language can be studied at different levels, of course. It has a biological basis—partly anatomical (the vocal tract, for instance), partly neurophysiological (neural patterns capable of controlling the articulatory gestures of word production)—but it is also controlled at a higher level by the rules of the language itself. Apparently, language is another supra-individual entity; it is not a private issue, as all language serves a social function. It enables many individuals to think together and work together as if they were a single super-individual. Undoubtedly, language is an inter-individual tool; in order to communicate and to be understood, we need to follow the rules of the language. A language has to be mastered so we can communicate with other users of the same language. In spite of all our good intentions, we may not always say what we mean (or mean what we say) because personal thoughts have to be conveyed by an interpersonal vehicle.

Language is not something you create on your own; if you developed your own private language, no one else would be able to understand you. Granted, individuals may add their own new words to their common language pool—provided they happen to catch on—but the language itself is not their own invention. It is a system different cultures have developed differently in order to categorize the world they live in. It is an emergent entity—not only a tool for expressing thoughts but also a system which shapes those thoughts in a downward causation.

The previous sentence might deserve more attention: Language plays a role not merely in spreading knowledge, but also in acquiring it. According to the Sapir-Whorf hypothesis, for instance, the structure of a language affects the ways in which its respective speakers conceptualize their world—which makes for some kind of world view—and thus shape their cognitive processes. The culture determines what is important to distinguish; Eskimos have many different words for different kinds of snow; Americans have many different words for various kinds of beer and whiskey. Distinctions depend on needs. That would be another case of downward causation. The said hypothesis has inspired many to think about how it might be possible to influence thought by consciously manipulating language.

There are many more issues regarding the reductionism-holism debate in the social sciences, but they are beyond the limited scope of this book. I will just end this chapter with one more question: How do social scientists actually do their research? As a matter of fact, they use a diversity of research methods, such as case studies, historical research, interviewing, participant observation, social network analysis, survey research, and statistical analysis. But even if they use direct observation of human behavior and its social interactions, they should never just pretend that humans cannot speak. Studying certain aspects of human behavior requires that humans can "speak their mind" in the language of their environment. A strict reductionist approach that precludes this kind of information may leave a lot behind and unanswered. In short, most social scientists would agree that there are biological, linguistic, historical, as well as cultural aspects to any problem they want to study.

For further reading:

Bochenski, Joseph M. *The Road to Understanding*. North Andover, MA: Genesis Publishing, 1996, chapter 8.

Dupré, J. *The Disorder of Things: Metaphysical Foundations of the Disunity of Science*. Cambridge, MA: Harvard University Press, 1993.

Lucy, John A. "Linguistic Relativity." *Annual Review of Anthropology*, 1997, 26: 291–312.

Woo, Henry K. H. "Scientific Reduction, Reductionism, and Metaphysical Reduction." *Methodus*, December 1990, 61-68.

16. Artificial Intelligence

The study of artificial intelligence (further abbreviated to *AI*) is sometimes called an applied or technological science, to distinguish it from so-called pure and academic science. However, this distinction may create the false impression that science and technology are very far apart from one another. In fact, hardly any theoretical understanding of nature has been gained without some technical intervention in nature and society. Besides, there is hardly any technology that does not presuppose a thorough theoretical understanding of nature and society.

Technology is usually related to techniques based on "pure" science, and conversely, "pure" science is often knowledge acquired by techniques. Moreover, "applied" investigation has often been an important source of new academic knowledge. It is probably no coincidence that many discoveries about the working of the human body were inspired by the latest technological contraptions of the time. Our understanding of many human organs came in the form of machine-like mechanisms. The camera with its lens helped us understand the working of the human eye. Bellows clarified how the lungs can do their work. Pumps revealed what the heart does for blood circulation. Computers threw some light on the working of the brain. Food is to the body what fuel does to the steam-engine.

No wonder then, when things go wrong, we use technological devices to correct them: lenses, hearing aids, pace makers, and prostheses. In all these examples, the case could be made that technology was an important driving force for scientific advancement. Scientific discoveries often start with technological inventions. The science of microbiology, for instance, originated largely from Pasteur's investigations of practical problems in the beer, wine, and silkworm industries. So, in a sense, they are all *explorative* sciences, and therefore, artificial intelligence deserves to be in this series of test cases as well.

To do justice to its "pure" as well as "applied" aspect, the field of AI could be defined as "both the study and the design of intelligent agents." John McCarthy, who coined the term AI in

1955, defines it as "the science and engineering of making intelligent machines." This makes AI an interdisciplinary field, in which a number of sciences and professions converge, including computer science, psychology, linguistics, philosophy, and neuroscience, as well as other specialized fields such as artificial psychology. The question is, though, what is meant by an "intelligent" agent or machine? A technical description would be that it is a system that perceives its environment and takes actions that maximize its chances of success—whatever these terms may stand for.

In order to do all of the above, an AI system must be more than a *one*-level phenomenon, to begin with. It is at least composed of *two* levels. As we discussed earlier with regard to computers (in chapter 13), the content at the top level is manifestly independent of its material substrate at the bottom level. No possible knowledge of the computer's materials can yield any information whatsoever about the actual content of its computations. In the words of George Gilder, "What technologists call the 'physical layer' of a computer network is the lowest level of silicon chips and silica fiber on the bottom, whereas the programs and content are at the top." So AI has a structure of at least two levels. It makes for an ideal case of a *systems* approach, because a computer in fact *is* as system in itself.

The idea that machines are like humans and that humans are like machines is not new but is probably best known from the French physician and philosopher Julien de La Mettrie who wrote a book entitled *Man a Machine* (1748). His ideas became part of a world-view called mechanicism. Mechanicism has since turned into an important philosophical doctrine to declare that all living objects, including human beings, are only and merely machine-like automata, which just follow all the physical laws of the universe, controlled by the machinery of their bodies. In this line of thought, Marvin Minsky, a pioneer in the field of artificial intelligence, described human beings as mere "machines made of meat." Undoubtedly, the machine metaphor has been extremely fruitful for the advancement of science. It has been said that science has profited more from the steam engine than the steam engine from science. As a matter of fact, most of the time, science is in search of "mechanisms." However, looking at an organism *as if* it were a machine does not *make* it a machine.

What all man-made machines, including AI systems, have in common is the fact that they follow the same rule: They are only useful because the laws of nature have been harnessed within certain set of constraints. It is telling that Michael Polanyi (see section III) uses the machine analogy to show that a specific construction represents only one of a very large number of possible arrangements of its parts, all of which are compatible with the laws of physics and chemistry. His thesis is that a machine based on the laws of physics is not explicable by the laws of physics alone, because the *structure* of a machine is what he calls a "boundary condition" extraneous to the process it delineates. These boundary conditions, instead of being random, are determined by a higher level whose properties are dependent on, yet distinct from, the lower level from which they emerge. Needless to say that the "intelligence" part of "artificial intelligence" is not located in its lowest level.

But AI claims go much further than what a personal computer, or even a robot, can do. The AI field was founded on the claim that a central property of humans—their intelligence—can be so precisely described that a machine can be built to simulate it. However, the idea that a machine can simulate human intelligence raises many philosophical issues, especially of what is meant by "human intelligence."

In general, intelligence works with perceiving sense-data and processing them. Many animals show some form of intelligence in their behavior, because intelligence is a brain feature and as such an important tool in survival. Therefore, animals can process images more or less intelligently. They show various forms of intelligence: We find spatial intelligence in pigeons and bats, social intelligence in wolves and monkeys, formal intelligence in apes and dolphins, practical intelligence in rats and ravens, to name just a few. In other words, intelligence is a matter of processing sense-data—something AI systems could do as well if they are outfitted with "sensors." Robots, for instance, can "cleverly" process sounds, images, and the like, and then react "intelligently."

Some take the claim of intelligence much further by posing the question as to whether AI machines can also *think* like humans do. A few of the most influential answers to this question are as follows. (1) Alan Turing reasoned that we need not decide if a machine can "think"; we need only decide if a machine can act as intelligently as a human being, which forms the basis of the Turing test. (2) The Dartmouth Project of 1955 stated that every aspect of learning, or any other feature of intelligence, can be so precisely described that a machine can be made to simulate it. (3) Another response is that general intelligent action is a matter of explicit symbolic knowledge which consists of formal operations on symbols. Any of these three answers would allow for artificial intelligence as a simulation of human intelligence by explicitly limiting what we mean by thinking and intelligent actions.

But the boundaries of the debate have gradually been extended. The question whether AI systems can *think* like humans has been stretched much farther into the domain of the human *mind*: Would an appropriately programmed computer with the right inputs and outputs just *simulate* a mind, or actually *have* a mind? The former part of the question, simulating a mind, is relatively neutral, but the latter part, having a mind, would have enormous philosophical repercussions, if it were true (see chapter 13).

The position of the latter part is often called "strong AI." It was already behind some of the statements of early AI researchers. For example, in 1955, AI founder Herbert Simon declared that we can explain now "how a system composed of matter can have the properties of mind." Cognitive scientist John Haugeland even wrote that "we are, at root, computers ourselves." More specifically, one could state that strong AI, in David Cole's words, represents the view that suitably programmed computers can understand natural language and actually have other mental capabilities similar to the humans whose abilities they mimic.

One of the leading attacks against these more extravagant claims comes from the Berkeley philosopher John Searle. He introduced what is widely known now as the "Chinese Room Argument"—first presented in 1983. Over the last three decades, this argument was the subject of many discussions.

In January 1990, the popular monthly *Scientific American* took the debate to a general scientific audience.

Searle himself summarized the Chinese Room argument as follows: Imagine a native English speaker who knows no Chinese locked in a room full of boxes of Chinese symbols (a data base) together with a book of instructions for manipulating the symbols (the program). Imagine that people outside the room send in Chinese symbols which, unknown to the person in the room, are questions in Chinese (the input). And imagine that by following the instructions in the program, the man in the room is able to pass out Chinese symbols which are correct answers to the questions (the output). Then, Searle asserts, this program enables the person in the room to pass the Turing Test for understanding Chinese, yet that person does not understand a word of Chinese.

The question Searle wants to answer is this: Does the machine literally "understand" Chinese? Or is it merely simulating the ability to understand Chinese? If you can carry on an intelligent conversation with an unknown partner, does this imply your statements are really "understood"? The position of strong AI would claim such is indeed the case. In contrast, Searle argues that what the machine is doing cannot be described as "understanding," and therefore, the machine does not have a "mind" in anything like the normal sense of the word. A computer can translate language, as several online translator tools show us, but the computer does not really understand language, let alone think of itself as an "I" who does the understanding. Hence, Searle concludes, strong AI is mistaken.

What Searle means by "understanding" is what some philosophers call "intentionality"—the property of being about something, of having content. It is the "mystery" of how mental thoughts about objects come to be *about* those objects. Thoughts are about something mental, about something beyond themselves. To use an analogy, anything that shows up on a computer monitor remains just an "empty" collection of "ones and zeros" that do not point beyond themselves until some kind of human interpretation gives sense and meaning to the code and interprets it as being "about" something. Think of what we call a picture: It may carry information, but the picture itself is just a piece of paper that makes "sense" only when human beings interpret the picture as being about something. The same with books: They provide lots of information for "book worms," but to real worms they only have paper to offer. That's where the need for intentionality and about-ness comes in. This raises the question whether a computer really "understands" what it is doing. It manipulates symbols or numbers that mean something to the human programmer, but do they mean anything to the computer? Does it "know" that the string of symbols it prints out refers to circles, for instance, rather than to melons or planets?

In the 19th Century, psychologist Franz Brentano re-introduced the term intentionality from scholastic philosophy and held that intentionality was the "mark of the mental." The term was also widely used by the philosopher Edmund Husserl. Then Gottlob Frege made it a standard practice in analytic philosophy to investigate the intentional structure of much human thought by inquiring into the logical structure of the language used by speakers to express it or to ascribe it to others. In other words, beliefs are taken as intentional states—which means they have propositional content, e.g. "one believes that p."

If this is true, understanding information must be more than merely making use of that information, otherwise we would end up with bizarre conclusions. The physicist Stephen Barr makes the following comparison: "An ordinary door lock has 'information' mechanically encoded within it that allows it to distinguish a key of the right shape from keys of other shapes. [...] Does the lock understand anything? Most sensible people would say not. The lock does not understand shapes any more than the fish understand shapes. Neither of them can understand a universal concept." As a matter of fact, universal concepts are uniquely human. Whereas the brain can handle signals and images, it seems that only the human mind can deal with concepts. Images can have some degree of generality—we can visualize a circle without imagining any specific size—whereas concepts have a universality that images can never have—the concept "circle" applies to every circle without exception. Because images are inherently ambiguous, open to various interpretations, we need concepts to give them a specific interpretation. That's how mental concepts transform "things" of the world into "objects" of knowledge, thus enabling humans to see with their "mental eyes" what no physical eyes could ever see.

It is to be expected that Searle's argument would come under attack from many sides. Without going into those discussions here, it could be stated that Searle's thought experiment appeals to our strong intuition that a person who does exactly what the computer does would not thereby come to understand Chinese. Most critics have held that Searle is quite right on this point—no matter how you program a computer, the computer will not literally have a mind, and the computer will not understand natural language but only process it. Is that a final defeat for strong AI? Probably not, but a final defeat of the Chinese Room Argument is yet to be seen. John McCarthy, who invented the term artificial intelligence, keeps claiming that "machines as simple as thermostats can be said to have beliefs."

There are still many philosophers and scientists who would maintain that a computer does not and cannot have a mind, and that our minds are not computers. Do they have any good arguments for their position? One of those arguments goes back to the legendary mathematician Kurt Gödel from Princeton University, who showed us that if we knew the program that a computer uses, we can in a certain sense outwit the computer. Based on this, the philosopher J. R. Lucas and the physicist R. Penrose argued that if we ourselves were just computers, we would be able to know our own programs and thus outwit ourselves. Since this is certainly not possible, we cannot equate ourselves with a computer.

This argument too has been attacked from many sides, one of them being that the human computer is not, as Gödel requires, consistent. But most of those attacks have ingeniously been refuted. One rebuttal is that humans do have the ability to reason consistently in a way that an inconsistent machine would not have and would not be able to fake. Besides, those who claim that humans carry inconsistent programs are very sure that the argument of Lucas and Penrose is wrong. Let's leave it at that.

Another argument against the strong version of AI is based on the distinction between human intelligence and human intellect. Intellect is arguably very different from intelligence (see chapter 13). Like intelligence, intellect also uses sense-data such as images and sounds, but unlike intelligence, it changes perception into *cognition* by using concepts and reasoning, and thus making sensorial experiences intelligible for the human mind. We could use Michael Augros's analogy again: You cannot count what you are seeing without using your eyes, but that does not mean your eyes are doing the counting. Similarly, it is clear that we cannot understand anything without using our brains, but it does not follow that our brains are doing the understanding. That's where the *intellect* comes in.

A concept may be as simple as a "circle" or as complex as a "gene," but a concept definitely goes beyond what the senses provide. Concepts are not perceived images. Images are by nature ambiguous, open to various interpretations; so we need concepts to interpret them. We do not "see" genes but have come to hypothesize and conceptualize them in a concept. We do not even see circles, for a "circle" is a highly abstract, idealized concept (with a radius and diameter). We can even conceptualize something that we cannot visualize—something like a circle with four dimensions, for instance. Once these concepts have been established and mastered, we have become "regular observers" of "circles" and "genes." But again, these are not images but concepts.

Mental concepts transform "things" of the world into "objects" of knowledge; they change experiences into observations, thus enabling humans to see with their "mental eyes" what no physical eyes could ever see before. To be sure, all we know about the world does come through our physical senses, but this is then processed by the immaterial intellect that extracts from sensory experiences that which is *intelligible* to us with the aid of concepts. Without a concept such as "gene," we do not understand genetics. Obviously the intellect may assist human intelligence, but it goes beyond it. Whereas intelligence can also work on its own, as it does in animals and even robots, these latter do not have intellect, because they lack mental concepts and conscious reasoning. So machines may have intelligence to various degrees, but this does not entail that they also have intellect like humans do.

Then there is at least one more reason that makes it very unlikely that a computer has a mind. Artificial intelligence is not just a two-level but in fact a *three*-level phenomenon. Above the level of the "hardware" and above the level of the "software" is the level of the designers of the AI system. A computer can be programmed to make decisions with an if-then-else structure, for instance. But this does not mean that the computer actually deliberates and decides—programmers did, and they make it look like the machine does too. The "reason" why a calculator adds or a pump pumps is that a machine was designed for that purpose—and not because it "intends" to.

The case could be made that without human subjects, computers just cannot "think," let alone have a "mind." Computers do not create thoughts, but they may, just like books and radios, carry thoughts that were created by the mind of a human subject—for example when using a word-processing program. Consider a voice recognition system; it does not really understand what it is programmed to "recognize." Cameras and other recording tools only record, but they do not interpret or understand what they record—we do.

In other words, computers only do what we, human beings with a mind, cause them to do, for we have proven to be champion machine builders. A computer may play chess better than Kasparov or any other champion, but it plays the game for the same "reason" a calculator calculates or a pump pumps—the reason being that it is a machine designed for that purpose. Besides, it can only play chess; to do other things as well, it would have to be programmed in an even more complicated way.

Because of this, all AI systems presume a designing mind, at a higher level of organization, or at least some form of artificial shaping, that creates the right boundary conditions for their performance. To put this in a catchphrase, not only is there "a machine in the man," but there is also "a man in the machine" who designed the machine. So the popular slogan "Man versus Machine" should actually read "Man versus Man"— versus the Man who built the machine. Are these considerations final and decisive? Probably not, but they cannot just be ignored either.

Perhaps one more consideration should be mentioned as to why there might be inherent limits to what machines can do. Roger Penrose is among those who claim that Gödel's theorem limits what machines are capable of doing. Kurt Gödel rigorously and mathematically proved in his famous so-called *incompleteness theorem* that coherent systems are incomplete (while complete systems are incoherent). As to our discussion, this would imply that no coherent system—not even the artificial intelligence of a machine—can be completely closed; any coherent system is essentially incomplete and needs additional "help" from outside the system, according to Gödel. That's where the machine designer might come in.

Of course, one could always raise the objection that we have no right to say what the technology of artificial intelligence will *never* achieve, because that would be an a priori answer to an empirical question. But again, those who consider that question empirical are plainly arguing a priori themselves. It may not be an empirical question but rather a philosophical one.

For further reading:

Berto, Francesco. *There's Something about Gödel: The Complete Guide to the Incompleteness Theorem*. Hoboken, NJ: John Wiley and Sons, 2010.

Penrose, Roger. The Emperor's New Mind: Concerning Computer, Minds and the Laws of Physics. Oxford University Press, 1989.

Scientific American, Volume 262, Number 1, January, 1990.

Searle, John. *The Rediscovery of Mind*. Cambridge, MA: MIT Press, 1992.

17. Philosophy of Science

This chapter does not really belong in the section of test cases regarding reductionism and holism in the sciences. Yet, in an analogous way, it would be possible to distinguish a "reductionist" as well as a "holistic" approach in certain areas of the philosophy of science. Let us see how and why.

One of the areas that seem to have a "reductionist" as well as a "holistic" version is the discussion as to how scientists develop scientific theories and hypotheses. The classical view of so-called inductivism could be called reductionist in a metaphorical sense. Inductivism is the traditional model of the scientific method explained in 1620 by Francis Bacon. According to the Baconian model, a scientist observes nature, then tentatively poses a modest axiom to generalize an observed pattern, then confirms it by many observations, next ventures a modestly broader axiom, and finally confirms that too, by adding more and more observations.

In essence, inductivism is a way of building theories up "from the bottom," from observation to observation—that is why it could be called a "reductionist" method. It is basically a form of "generalizing induction," which takes us from a general statement about "some" instances to a universal statement about "all" instances. One starts with singular statements about events or instances, and by adding more and more instances, one ends up with a universal statement. That is a way of doing research that seems to have worked very well in science.

However, the fact that induction has worked in the past does not indicate that it will work in the future. For this reason, it is hard to justify induction without using inductive reasoning. Defending inductive reasoning based on the fact that it has worked well so far is itself a form of inductive reasoning and therefore makes for a circular argument. Science understood as an inductive enterprise cannot be proven inductively by using empirical evidence, and thus science cannot be proven scientifically. Yet, induction is considered a very popular method in science.

The classical "heroes" of induction are Francis Bacon and John Stuart Mill. The former believed he had found a method stripped of all philosophical fancies. The latter believed he could offer us an explicit set of clean and safe inductive rules which would help us in a "mechanical" way to seek and find the *cause* of something—"from the bottom up," so to speak. The first of Mill's rules, for instance, tells us exactly how to find the cause of something. Its recipe is as follows: Look for all the circumstances preceding some phenomenon and find out which circumstance in particular occurs every time the phenomenon takes place. In finding that particular circumstance, we have demonstrated by induction that in the many cases studied, there is one particular circumstance which is the *cause* of the phenomenon in question. Thus, we would end up with a universal statement to the effect that circumstance X is the cause of phenomenon Y in *all* cases. If a certain sickness, for instance, occurs in human beings who all carry a certain type of bacteria, we must assume that it is this kind of bacteria that causes the sickness.

However, some have argued there are many problems with a reductionist approach like this. To begin with, Mill's rules cannot be applied as simply and mechanically as he thought. After all, many things can go wrong when applying his rules. First of all, we have to be aware of the old saying "After this and yet not because of this." Sunrise may be after cockcrow, but it is not caused by cockcrow. In a similar way, the sickness in question may occur after a bacterial infection, but without being caused by it. Then there is the problem that the actual cause of the sickness may be completely different from the one we had come up with—for example, a viral infection instead which only has a better chance of developing after a certain bacterial infection. And finally, there is the problem that there could be more than one cause involved. A poor immunity system, for example, may add to the chances of a bacterial or viral infection.

This leads us to a second more general and basic problem regarding generalizing induction. The act of generalizing is based on the fact that objects and events are similar and look alike in certain respects—e.g. "being infectious," "being toxic," "being genetic," and so on—but this similarity is not visible until we know already what it is that "similar" cases have in common, which requires the proper concepts. We need to identify first what is relevant to our problem, because similarity cannot be established until it has been identified in a word, or actually in a concept. We cannot mechanically infer from a few cases to all similar cases until their similarity has been conceptualized first. Since things can be "alike" in many, many ways, we need a unifying concept of similarity first before we can classify and categorize things as similar. Before we can "notice" a carnivore, we need the "notion" of a carnivore to begin with.

In our cases, the similarity stands or falls with concepts such as "infection," "toxicity," and "gene." Without cognition there is no re-cognition. Hence, there is always a conceptual leap involved. Take our case again of a sickness caused by bacterial infections. For the sickness to be explained that way, somebody has to come up with the idea or hypothesis of a bacterial infection. Before Robert Koch and Louis Pasteur had published their experiments, no one would ever have thought human sickness could be caused by bacteria. Actually, the problem is that infinitely many factors may qualify as the potential cause of a certain phenomenon. Mill's rules of induction only work when we have before us *all* and *only* the facts relevant to the solution of our problem. But that is quite an assumption; most of the time we do not!

The following trivial example may clarify this in a different way. Imagine, you want to find out whether the headache you developed is caused by "gin on the rocks," or by "whisky on the rocks," or perhaps by "rum on the rocks." According to Mill's rules, you should extensively experiment with these three drinks... to find out that your headache is caused by ice cubes, because that is what all these drinks have in common. This conclusion seems quite reasonable, until you come to know of the more-embracing concept of "alcohol"—being a generic term for gin, whisky, and rum together. To arrive at a new concept, a mental leap is needed. Not any kind of inductive rule can achieve this for us. Searching is a matter of imagination rather than calculation. In the search phase there is more need of provisional ideas than of logical tools.

An added, even more serious problem of inductivism is that reasoning from observation to observation assumes that observations are clear-cut, preset elements underlying scientific knowledge and explanation. This assumption has been seriously questioned, though. Karl Popper used to say that the command "Observe!" does not make any sense, since no one would know what to observe. His point is that scientific theories just do not and cannot spontaneously emerge from observation. We do not "have" observations—like we have sensorial experiences—but we "make" observations. Philosophical giants such as Aristotle and Aquinas would put it this way: All we know about the world comes through our senses but this is then processed by the intellect that extracts from sensory experiences that which is intelligible. That is the reason why a camera cannot make observations. Cameras do not observe but they record. A surveillance camera, for instance, automatically "observes" every single detail because it does not know what to observe. That is why cameras cannot replace scientists—they may help them but cannot replace them.

The problem with any kind of images or pictures is that they do not show us observations until we give some interpretation to the things and events we see on the picture. As said earlier, it is through mental concepts that we transform "things" of the world into "objects" of knowledge; concepts change experiences into observations, thus enabling humans to see with their "mental eyes" what no physical eyes could ever see before. To be sure, all we know about the world does come through our physical senses, but this is then processed by the immaterial intellect that extracts from sensory experiences that which is intelligible in conceptual terms.

What could be concluded from this? It is highly unlikely that there are simple "rules of induction" by which hypotheses and theories can be mechanically generated from empirical data—let alone from so-called observations. Logic provides us with formal means that allow us to check an argument after it has been made. The rules of logic are not rules of discovery but at best rules of validation. Logic does not allow us to detect statements before the event, but only to check them after the event and provide them with a seal of approval, if possible. Hence, most philosophers would agree there is hardly any useful logic of discovery in science.

What would be the solution then, if we cannot build up theories by going from observation to observation? We could start at the other end, at a higher level where hypotheses and theories reside. This makes most philosophers of science maintain that scientific hypotheses are "happy guesses," which are not derived or discovered but they are invented first by a scientist. Research is first of all a matter of asking the right questions—by means of new concepts, models, hypotheses, and theories. The best way to search is to have an idea of what you are looking for. The search phase thrives on ideas; without ideas in the search phase, science would be blind. There is no such thing as seeing-in-a-neutral-way, or observing-without-expectation. The models we spoke about at the beginning (see chapter 1) are basically tools guiding us as to how and what to observe.

Examples of this approach can be found anywhere in science. Let me just use the simple case of how William Harvey discovered the principle—or model, if you will—of a closed blood circulation. His new idea was not discovered but invented. Harvey never saw the connecting blood capillaries needed for a closed circulation. Instead Harvey came up with an idea, a hypothesis, and a model. We do not know what exactly gave him this idea. Was it the thought of an outlet needed for a pumped inflow, or rather Aristotle's idea of perfect circular motion, or perhaps the developing technology of pumps? Whatever it was that guided him, Harvey hypothesized two one-way circulations controlled by a heart pump. Apparently, the capillaries did not show the closed circulation of blood, but it was the other way around: The very theory of a closed blood circulation would make the vessels visible when better microscopes later became available. Harvey's new theory was initially one more of invention than discovery.

In science, discoveries typically start as inventions—usually called hypotheses. However, not all inventions lead to discoveries. To use an analogy, the person who invented "Atlantis" did not discover Atlantis; it remains a legendary island until further notice. The same in science: Most inventions do not make it to the stage of discoveries. Yet some scientists think they have made a discovery when all they have in mind is an invention, a hypothesis. However, a hypothesis is only an invention in the mind until it has been proven to be also a discovery in reality. Peter Medawar's wise advice to a (young) scientist is that "the intensity of the conviction that a hypothesis is true has no bearing on whether it is true or not." What is needed is a rigorous test of whether the hypothesis holds after we derive test implications from it. Karl Popper put it this way: "Every 'good' scientific theory [...] forbids certain things to happen." Albert Einstein said something similar: "No amount of experimentation can ever prove me right; a single experiment can prove me wrong." But all of this assumes we do have a hypothesis in mind. Hence, scientists should always be ready to take "no" for an answer if falsifying evidence points that way (but there is more to it, of course).

The approach used here seems to be somehow "holistic." Research is like a downward causation process leading from hypotheses to test implications, followed by confirming or falsifying observations. Observation is seen as a "theory-laden" phenomenon that does not originate from "barren facts at the bottom" but from "scientific imagination at the top." From here it is only a small step to the holism of what has become known as the Duhem-Quine thesis, which states that hypotheses are not tested in isolation but only as part of whole bodies of theory. Pure induction, utterly free from expectations, is arguably just an illusion. To put it in more general terms, a science free from assumptions and commitments may simply not exist. If so, reductionism would have to yield to holism in this discussion.

We find a similar discussion—in terms of reductionism and holism—regarding the issue of causality in science. Scientists assume there is some kind of order in this universe, based on the rule that says, "Like causes produce like effects." This is a vital rule in science. If there were no order in the universe, then it would make no sense to search for laws in physics, chemistry, biology, or the social sciences. Only because there is "law and order" are scientists able to explain and predict—which would not be possible in a world of disorder and irregularity. It is only because we may assume that like causes produce like effects that we are able to explain and predict in science.

However, this rather solid basis of the scientific building was eaten away by the "reductionist" approach of the philosopher

David Hume. Ironically, Hume found multiple obstacles to Bacon's reductionist kind of inductivism, but in turn, he created his own version of reductionism. He stressed that generalizing induction—that is, unrestricted generalization from some particular instances to all instances, by stating a universal law—is illogical. His reason for stating this was that humans observe only sequences of sensory events, not something like cause and effect. According to Hume, we do not actually experience the necessary connection between events—we only observe a constant conjunction between two events. Hume argued that the "mechanism" of causality is merely a kind of illusion produced by habit or custom: We simply *imagine* causality. In his own words, "From causes which appear similar we *expect* similar effects. This is the sum of our experimental conclusions."

This stand made Hume one of the first skeptic philosophers, who questioned the very idea of causality—actually of objective truth in general. He argued that all connections we observe in life are nothing but constant conjunctions in our minds, and our perceptions of them never give us insight into the modus operandi of the connection. So he declared causal connections to be mere "metaphysical" inventions, based on an illusion. As a consequence, causal connections in themselves are ultimately subjective phenomena on Hume's view.

Many have criticized Hume's analysis since. First of all, this analysis would erase the important scientific distinction between causality and correlation; they would both be reduced to a series of mere subjective associations, conjunctions, or generalizations. In contrast, scientists always want to make sure that causality is not merely a matter of correlation. Second, although we do know the world through sensations or sense impressions, many would counter these are just the means that give us access to reality. There is a way things are, independent of how they may be apprehended. The philosopher John Haldane put it this way, "One only knows about cats and dogs through sensations, but they are not themselves sensations, any more than the players in a televised football game are color patterns on a flat screen." Knowledge does rest on sensation, but this does not mean it is confined to it. It is "reality" that sometimes forces us to revise our theories.

If we were to follow Hume's philosophy, we would end up with what the late physicist and historian of science Stanley Jaki calls "bricks without mortar." The "bricks" may be the result of reductionism, but the "mortar" has to come from holism. Jaki says about Hume's sensations, "the bricks he used for construction were sensory impressions. Merely stacking bricks together never produces an edifice, let alone an edifice that is supposed to be the reasoned edifice of knowledge." These "bricks" are reductionist elements that need the "mortar" of holism.

Nonetheless, skeptics like Hume find a flaw in every truth claimed, even the truth of causality. Skepticism makes for a very restrained view on the world—actually so restrained that absolute skeptics cannot even know whether they have a mind to doubt with. Skeptics seem to turn things the wrong way. Granted we often do need to eliminate errors in order to get to the truth, yet our ultimate goal is not to avoid errors, but to gain truth—to know rather than to know what we do *not* know. We want to know, not to know what we do *not* know. Skeptics, on the other hand, make it their final goal to avoid errors, in denial of the fact that eliminating errors is only a means to gaining truth about reality. Skepticism is at best a method to avoid errors, but it never leads to truth. Once we begin questioning the trustworthiness of our brains and senses and intellect and reason, there is no way of establishing their trustworthiness again independently of trusting them.

In short, Hume leaves us in a cognitive desert. It is a "reductionist" approach that calls for a more "holistic" approach. Causality is not something that arises from individual cause-and-effect associations, but it is a rational assumption in the human mind that makes science—and all other kinds of knowledge, for that matter—possible. It is not a scientific discovery based on a series of individual cases. In other words, it is not an intra- but extra-scientific notion; it is in fact a proto-scientific notion that must come first before science can even get started.

Because of this, we should not confuse general, inductive statements in science such as "All iron expands with heat" with universal principles in philosophy. The former are confirmed by testing more and more instances of iron under various temperature conditions; such cases are examples of inductive generalization. In contrast, universal principles such as "All expanding of iron has a *cause*" are true, independently of any particular cases; their truth does not increase by testing more and more instances, because their truth does not depend on inductive generalization. Science cannot explain causality itself but must assume it for its explanations. It is not an *a priori* form of thought but a *given* in reality, "engraved" in the structure of the universe. It is a given that enables intelligibility, allowing us to grasp reality. That's how we can derive the law that all iron expands with heat from higher level laws in physics.

No wonder then, science can never prove there is order in this universe, but instead must *assume* it as a universal principle before it can prove anything. As Dimitri Mendeleev said, "It is the function of science to discover the existence of a general reign of order in nature and to find the causes governing this order." Even the tool of falsification, for instance, is essentially based on this very assumption of order: The fact that scientific evidence can refute a scientific hypothesis is only possible if there is in fact causality, order, and lawfulness in this universe to begin with. Without "law and order" in nature, there simply could not be any falsifying evidence. When we do find falsifying evidence, we do not take such a finding as proof that the universe is not orderly, but instead as an indication that there is something wrong with the specific order we had conjectured up in our minds.

Apparently, falsification itself is based on order and cannot be falsified by disorder. Hence, counter-evidence does allow us to falsify theories, but not the principle of falsification itself. In utter amazement, Albert Einstein once wrote in one of his letters, "But surely, a priori, one should expect the world to be chaotic, not to be grasped by thought in any way." Einstein was enough of a philosopher to realize the importance of a given order, one of the main pillars of science. There must be "something" in reality that makes the universe law-abiding.

Some scientists may consider this kind of reasoning a despicable form of metaphysics. They have come to take "metaphysics" as a "dirty" word anyway, but the fact is that no one can live without metaphysics. Those who reject metaphysics are in fact committing their own version of metaphysics. Rejecting metaphysics can only be done on metaphysical grounds, for any rejection of metaphysics is based on a metaphysical viewpoint regarding what the world "really" is like. Metaphysics may be a "dirty word" to some, but all of us are surrounded by it.

So what is the relationship then between science and metaphysics? Let me put this in a catchphrase: Is physics the basis of our meta-physics, or is our meta-physics the basis of physics? The former viewpoint is rather popular among scientists such as Sigmund Freud who believed himself to be free of any worldview. They proclaim that there is no worldview, no metaphysics in what they claim—at best, their metaphysics can just be "reduced" to physics.

The more technical term for this is *scientism*. Supporters of scientism claim that science provides the only valid way of finding truth. They pretend that *all* our questions have a scientific answer phrased in terms of particles, quantities, and equations. Their claim is that there is no other point of view than the "scientific" point of view. They believe there is no corner of the universe, no dimension of reality, no feature of human existence beyond its reach. In other words, they have a dogmatic, unshakable belief in the omni-competence of science. They portray scientists as a bunch of emotion-free, white-coated people who battle collectively to snatch secrets from the stubborn universe—mostly and preferably in a reductionist way.

A first reason for questioning the viewpoint of scientism is a very simple objection: Those who defend scientism seem to be unaware of the fact that scientism itself does not follow its own rule. How could science ever prove all by itself that science is the only way of finding truth? There is no experiment that could do the trick. Science cannot pull itself up by its own bootstraps—any more than an electric generator could run on its own power. So the truth of the statement "no statements are true unless they can be proven scientifically" cannot itself be proven scientifically. It is not a scientific discovery but at best a philosophical or metaphysical stance or dogma. It declares everything outside science as a despicable form of metaphysics, in defiance of the fact that all those who reject metaphysics are in fact committing their own version of metaphysics.

A second reason for rejecting scientism is that a successful method like the one science provides does not automatically disqualify any other methods. Scientism poses a claim that can only be made from outside the scientific realm, thus grossly overstepping the boundaries of science. So if it is true, it becomes false. It steps outside science to claim that there is nothing outside science and that there is no other point of view—which does not seem to be a very scientific move. First it limits itself to what science can investigate, and then it claims there is nothing else left. The late UCB philosopher of science Paul Feyerabend came to the opposite conclusion when he said that "science should be taught as one view among many and not as the one and only road to truth and reality." The late British analytical philosopher Gilbert Ryle phrased this idea in his own terminology: "the nuclear physicist, the theologian, the historian, the lyric poet and the man in the street produce very different, yet compatible and even complementary pictures of one and the same 'world.'"

A third reason for questioning scientism is the following. Scientific knowledge does not even qualify as a superior form of knowledge; it may be more easily testable than other kinds, but it is also very restricted and therefore requires additional forms of knowledge. Mathematical knowledge, for instance, is the most secure form of knowledge but it is basically about nothing. Consider the analogy used by the philosopher Edward Feser: A metal detector is a perfect tool to locate metals, but there is more to this world than metals. That is exactly where scientism goes wrong: Instead of letting reality determine which techniques are appropriate for which parts of reality, scientism lets its favorite technique dictate what is considered "real" in life—in denial of the fact that science has purchased success at the cost of limiting its ambition.

To characterize the dogma of scientism, the late psychologist Abraham Maslow used the following analogy: If you only have a hammer, every problem begins to look like a nail. So we should be careful not to idolize our "scientific hammer," because not everything is a "nail." Even if we were to agree that the scientific method gives us better testable results than other sources of knowledge, this would not entitle us to claim that only the scientific method gives us genuine knowledge of reality. No wonder this has led some to criticize scientism as a form of circular reasoning. The late philosopher Ralph Barton Perry expressed this as follows: "A certain type of method is accredited by its applicability to a certain type of fact; and this type of fact, in turn, is accredited by its lending itself to a certain type of method."

A fourth argument against scientism is that science is about material things, yet it requires immaterial things such as logic and mathematics. If logic is just a movement in the brain of a bewildered ape, good logic should be as misleading as bad logic. Logic and mathematics are not physical, and therefore not testable by the natural sciences—and yet they cannot be denied by science. In fact, science heavily relies on logic and mathematics to interpret the data that scientific observation and experimentation provide. Logic and reason are a perfect example of the kinds of immaterial phenomena that we all know exist, but that cannot be measured or analyzed by naturalistic science. Yet, these immaterial things are real and self-evident, even though they are outside of scientific observation. When we talk about proofs, we usually have mathematics in mind, but even mathematics has to start somewhere, and those starting points are called axioms—they are unprovable.

A fifth reason for rejecting scientism is that no science, not even physics, is able to declare itself a superior form of knowledge. Some scientists may argue, for example, that physics always has the last word in observation, for the observers themselves are physical. But why not say then that psychology always has the last word, because these observers are interesting psychological objects as well? Neither statement makes sense; observers are neither physical nor psychological, but they can indeed be studied from a physical, biological, psychological, or statistical viewpoint—which is an entirely different matter.

Often scientism results from hyper-specialized training coupled with a lack of exposure to other disciplines and methods. The findings of science are always fragmentary. There is no science of "all there is." Someday there may be a "Grand Unified Theory" (GUT) in physics," but that does not entail there is also a "Grand Unified Theory of Everything" (see chapter 5). What entitles us to claim that physics is all there is? Limiting oneself exclusively to a particular viewpoint such as physics is in itself at best a methodical, or even metaphysical, decision. However, to quote Shakespeare, "There are more things in heaven and earth, Horatio, than are dreamt of in your philosophy."

If the aforementioned arguments are valid, it is hard to believe, let alone to defend, that physics is the basis of metaphysics. It seems more warranted to take the "holistic" position that metaphysics is the basis of physics, and of all the other sciences. We found already some form of metaphysics in science—actually the kind of metaphysics that makes science possible in the first place. It includes various metaphysical assumptions about the existence of causality, functionality, emotionality, and cognitivity. It assumes that this universe is intelligible for us, and that it is a universe of "law and order." It is even a metaphysical idea that there is an organizational hierarchy of levels (see chapter 2). And then there is ontological reductionism and ontological holism, which are both metaphysical positions that determine what the basic elements in this universe are supposed to be. Because of all of the above, science could be called a metaphysics-based or faith-based enterprise, for even in science, it takes faith to study and understand this Universe. Science is something you need to believe in before you can practice it. It is only because of their trusting that nature is law-abiding and intelligible in principle that scientists have reason to trust their own scientific reasoning.

The physicist Richard Feynman is often quoted as saying, "Philosophy of science is about as useful to scientists as ornithology is to birds." This statement might be taken as a final verdict on the uselessness of what we tried to do in this book. But Feynman's comparison falls short and should not be taken too seriously, given the fact that the worthlessness of ornithology for birds cannot be blamed on the inadequacy of ornithology but rather on the inability of birds to grasp ornithology. Yet, I don't think this is something Feynman intended to say about the ability of scientists to learn from the philosophy of science. Perhaps Albert Einstein was right after all when he said: "It has often been said, and certainly not without justification, that the man of science is a poor philosopher."

Arguably science can learn something from philosophy, for the simple reason that there is no such thing as a strictly scientific level of disagreement, as distinct from a philosophical one. The two are intricately entwined. Even the reductionism-holism debate can only be understood in this context. Let me quote Einstein one more time when he focused on one branch of philosophy, epistemology, by saying, "Epistemology without contact with science becomes an empty scheme. Science without epistemology is—insofar as it is thinkable at all—primitive and muddled."

For further reading:

Crick, Francis. *What Mad Pursuit: A Personal View of Scientific Discovery*. New York, NY: Basic Books, 1988.

Haldane, John. "Hume's Destructive Genius." *First Things*, 218 (2011): 23-25.

Medawar, P. *Advice to a Young Scientist*. New York, NY: Basic Books, 1979.

Verschuuren, Gerard M. *Investigating the Life Sciences*. Oxford, UK: Pergamon Press, 1986.

V.   Index

About the Author

 Dr. Gerard M. Verschuuren is a human geneticist who also earned a doctorate in the philosophy of science. He studied and worked at universities in Europe and the United States. Currently semi-retired. he spends most of his time as a writer, speaker, and consultant.

Some of his more recent books are:

- *Investigating the Life Sciences* (Series: Foundations and Philosophy of Science and Technology. Oxford: Pergamon Press, 1986).
- *Life Scientists—Their Convictions, Their Activities, and Their Values* (North Andover, MA: Genesis Publishing Company, 1996).
- *Darwin's Philosophical Legacy—The Good and the Not-So-Good* (Lanham, MD: Lexington Books, 2012).
- *What Makes You Tick?—A New Paradigm for Neuroscience* (Antioch, CA: Solas Press, 2012).
- *The Destiny of the Universe—In Pursuit of the Great Unknown* (Saint Paul, MN: Paragon House, 2014).
- *It's All in the Genes!—Really?* (Charlestown, SC: CreateSpace, 2014).
- *Life's Journey—A Guide from Conception to Natural Death* (Kettering, OH: Angelico Press, 2016).

For more info: http://en.wikipedia.org/wiki/Gerard_Verschuuren.

He can be contacted at www.where-do-we-come-from.com.